THE CHEMISTRY AND
FERTILITY OF SEA WATERS

THE
CHEMISTRY AND FERTILITY
OF
SEA WATERS

BY

H. W. HARVEY
C.B.E., Sc.D., F.R.S.

'To all those creatures who
live in the sea'

CAMBRIDGE
AT THE UNIVERSITY PRESS
1969

PUBLISHED BY

THE SYNDICS OF THE CAMBRIDGE UNIVERSITY PRESS

Bentley House 200 Euston Road, London, N.W.1
American Branch: 32 East 57th Street, New York N.Y., 10022

Standard Book Number 521 05225 4

First Edition 1955
Second Edition 1957
Reprinted 1960
1963
1966
1969

First printed in Great Britain at the University Press, Cambridge
Reprinted by photolithography Unwin Brothers Limited Woking and London

CONTENTS

47149

PART II

THE CHEMISTRY OF SEA WATER

ACKNOWLEDGEMENTS

I am indebted to many friends and colleagues for suggestions and for reading parts of the manuscript of this book, to Mr F. A. J. Armstrong for collaboration in the final chapter, and to Miss L. M. Serpell for checking the bibliography.

References are, as far as possible, limited to the more recent papers dealing with the various subjects.

H.W.H.

MARINE BIOLOGICAL ASSOCIATION
PLYMOUTH, 1954

PREFACE TO THE SECOND EDITION

This edition is produced by photo-lithography. A few minor corrections are made in the text. New matter is added at pp. 217 ff.; an asterisk in the margin of the main body of the text refers the reader to this addendum at the end.

H.W.H.

APRIL 1959
Plymstock, Devon

PART I

THE ENVIRONMENT OF THE FLORA
AND FAUNA

CHAPTER I

COMPOSITION OF SEA WATER

GENERAL ACCOUNT

THE salt content of water from the open oceans, away from the immediate influence of melting ice or rivers, rarely exceeds 3·8 % and is rarely less than 3·3 %. Nine kinds of ions constitute $99\frac{1}{2}$ % of the salts in solution; these are found in remarkably constant proportion, the one to the other. On the other hand, some of the minor constituents and dissolved gases are found in widely differing proportion, due to the activity of living organisms. The organisms concentrate some elements in their tissues and body fluids; they adsorb other elements on their surface.

The salt content of sea water is usually expressed as its *salinity* or S‰, a convention which approximates to the weight in grams of dry salts contained in 1000 g. of the sea water. This value is obtained by titration with silver nitrate (p. 126).

Knowing the salinity, the concentration in grams per kilo of the nine major constituents—the 'conservative' constituents—can be readily calculated from a table such as table 1, derived from the mean values of many analyses.

The concentration of salts in this 35·00 S‰ water is 0·558 molar, and of these salts four-fifths behave as if dissociated, as indicated by the freezing-point, − 1·90° C. Hence the water will be isotonic with a 1·0 molar solution of an undissociated compound.

From the salinity of the water and its temperature, its specific gravity can be calculated (p. 130 and fig. 58). This allows its composition in terms of grams per litre to be derived.

If, in addition to salinity and temperature, the hydrogen-ion concentration is also known, its content of total CO_2, partial pressure of molecular CO_2 in solution, content of bicarbonate and of carbonate, can be calculated from the results of a long series of investigations by Buch.

For a natural sea water, to which no acid or alkali has been added, having a salinity of 35‰, table 2 shows the content of total CO_2 and the partial pressure exerted by the undissociated CO_2 in solution at 16° C. and at the hydrogen-ion concentrations shown.

TABLE 1. *The constituents in solution in an ocean water having a salinity* (S‰ = 35·00)

	Grams per kilo	Grams per litre at 20° C. (specific gravity 1·025)
Total salts	35·1	36·0
Sodium	10·77	11·1
Magnesium	1·30	1·33
Calcium	0·409	0·42
Potassium	0·388	0·39
Strontium	0·010	0·01
Chloride	19·37	19·8
Sulphate as SO_4	2·71	2·76
Bromide	0·065	0·066
Boric acid as H_3BO_3	0·026	0·026
Carbon: Present as bicarbonate, carbonate and molecular carbon dioxide	0·023 g. at pH 8·4 0·025 g. at pH 8·2 0·026 g. at pH 8·0 0·027 g. at pH 7·8	
As dissolved organic matter	0·001–0·0025 g.	
Oxygen (where in equilibrium with the atmosphere at 15° C.)	0·008 g. = 5·8 cm.³ per l.	
Nitrogen (where in equilibrium with the atmosphere at 15° C.)	0·013 g. = 10·5 cm.³ N_2 per l. + 0·28 cm.³ argon etc.	
Other elements	0·005	

The values depend upon pH, salinity and temperature. The total CO_2 in solution, at a particular pH, changes with change in salinity, nearly in direct proportion, between 27 and 38‰ S.

The total CO_2 in solution, at a particular pH, decreases about 1 % per rise of 1° C., and the partial pressure of CO_2 in solution at a particular pH increases about 1 % per rise of 1° C.

Table 2 allows rough estimates to be made for waters having salinities between 27 and 38‰ and temperatures between 6 and 26° C. Very precise estimates are possible, both for natural waters and for waters to which acid or alkali has been added, by means of the constants derived by Buch (pp. 153–182).

TABLE 2. *Content of CO_2 and its partial pressure in natural sea water at 16° C. and 35‰ S*

pH ...	7·4	7·8	8·0	8·1	8·2	8·3	8·4	8·5
Total CO_2 (mols per litre × 10^{-5})	238	235	216	211	206	199	191	183
Partial pressure of CO_2 (atmospheres × 10^{-5})	230	85	50	39	29	22	15	12

If this 35‰ S water at 15° C. is brought into equilibrium with the atmosphere, it will attain a pH of *c*. 8·16 and contain 45·4 cm³ CO_2 per litre present as bicarbonate and carbonate together with 0·26 cm³ of undissociated molecular CO_2 in solution and a very small proportion of undissociated H_2CO_3. The molecular CO_2 will then exert a partial pressure equal to that of the air, which contains 0·032–0·033 % of CO_2.

If plants grow in a litre of this water at 15° C. and abstract from it 2·24 cm³ CO_2 (0·1 millimol containing 1·2 mg. C)

(i) its pH will rise from 8·16 to 8·31;

(ii) the molecular CO_2 in solution will decrease from 0·26 to 0·17 cm³, and the partial pressure which it exerts in solution will decrease from 33 to 21 × 10^{-5} atmosphere;

(iii) some of the bicarbonate ions will change to carbonate ions, providing the 2·14 cm³ which the plants have consumed in addition to the 0·1 cm³ obtained from the decrease in molecular CO_2:

$$2NaHCO_3 \rightarrow Na_2CO_3 + H_2O + CO_2 \text{ abstracted by the plants.}$$

As a result of this growth, causing a rise of 0·15 pH, the plants will have synthesized organic matter containing 1·2 mg. carbon. This quantity is in addition to the quantity which has been synthesized and broken down in respiration by the plant cells during the growth period.

If animals in a litre of this water at 15° C. respire 2·24 cm.³ CO_2 into the water

(i) its pH will fall from 8·16 to 7·98;

(ii) the molecular CO_2 in solution will increase from 0·26 to 0·4 cm.³, and its partial pressure from 32 to 53×10^{-5} atmosphere;

(iii) the remainder of the CO_2 [2·24 – (0·4 + 0·26)] cm.³ will react with carbonate, producing bicarbonate:

$$Na_2CO_3 + H_2O + CO_2 \text{ produced by animals} \rightarrow 2NaHCO_3.$$

As a result of this respiration, causing a fall in pH of 0·18, the animals will have entirely oxidized, in their respiratory activities, organic matter containing c. 1·2 mg. C.

As respiration by the animals proceeds, the fall in pH and the rise in partial pressure both become progressively greater for each unit of CO_2 respired.

Temperature has an effect upon the system. Thus, if this 35‰ S water is cooled from 15 to 5° C. and again brought into equilibrium with the atmosphere, its pH will have changed from 8·16 to c. 8·12. The molecular CO_2 will have risen from 0·26 to c. 0·36 cm.³ per litre and the CO_2 present as carbonate and bicarbonate from 46·5 to c. 47·6 cm.³ per litre.

When a strong acid is added to the water, first carbonates and then bicarbonates are decomposed, setting free molecular CO_2, of which the partial pressure rises rapidly. The addition of 2·4 cm.³ N HCl is just sufficient to decompose all the carbonate and bicarbonate in a litre of 35‰ S water.

Sea water of around 35‰ S in equilibrium with the atmosphere is saturated or supersaturated with respect to $CaCO_3$; on boiling sea water a mixture of $Mg(OH)_2$ and $CaCO_3$ is precipitated; on adding an alkaline hydroxide, $Mg(OH)_2$ is precipitated.

ORGANIC SUBSTANCES IN SOLUTION

Dissolved organic substances have been estimated in sea water from the North Atlantic in quantities of about 5–6 mg. per litre, and rather more in the Black Sea. These contain some organic nitrogen and phosphorus, which are in course of time set free, by bacteria or hydrolysis, as ammonia and phosphate. If the ammonia set free escapes being absorbed and built up into organic compounds by plants, it is oxidized to nitrite and then to nitrate through the agency of bacteria. Analogy with lake water suggests that the dissolved organic matter contains proteins, polypeptides and many amino acids, also traces of thiamine and biotin and vitamin B_{12}. Traces of some of these and of similar compounds may account for the difference in growth of certain species of both plants and animals, in waters collected from different water masses.

MICROCONSTITUENTS

In addition to the elements already listed, sea water contains in solution traces of many others whose concentration has been determined (p. 140). In all, these traces together rarely exceed 5 mg. per litre, most of which is fluoride and silicate. Very deep waters, which rise to the surface in the Antarctic, contain silicon to the extent of about 3·3 mg. per litre in the form of silicate. In the upper layers of water in tropic and temperate seas, absorption and utilization by diatoms reduce its concentration to as little as 0·001 mg. per litre.

Many of the other 'microelements' are concentrated several hundredfold or thousandfold in marine plants or animals; hence their concentration in different sea waters depends upon the previous biological history of the waters.

As far as is yet known, naturally occurring variations in the concentration of only *four* of these microelements affect marine plant life and as a consequence the animal population. Of others which are essential constituents of various species of living organisms (as Zn, Co, Cu, V, I, Mo and As), there appears to be no lack for the requirements of the plant and animal populations.

The supply of phosphate and of nitrogen, in forms available to the plants, limits their annual production over wide areas of the seas. This is a major factor, varying from area to area, which exerts a control upon the population of plants and in consequence upon that of the animals which depend upon them.

There is reason to suppose that the very low concentration of iron and manganese in the upper layers in some areas far from land may depress the growth rate and limit the plant population (p. 98). At all events the addition of minute quantities of these elements has a marked effect upon growth *in vitro*.

The very variable concentrations of these four microelements found in waters from Atlantic sources are shown in table 3, which is included in order to give a generalized picture of the quantities met with. They tend towards higher concentrations in the Pacific Ocean, in common with many of the other microelements.

TABLE 3. *Concentrations of microelements in Atlantic waters*

	Micrograms per litre	Microgram-atoms per litre
Phosphorus as inorganic phosphate in solution		
Antarctic:		
Upper layers	62–64	2
Below 150 m.	75–85	2·9–2·8
North Atlantic:		
Upper 20–50 m.	often < 1 in summer	0·03
Below 600 m.	*c.* 40	1·3
English Channel:		
Upper layers in summer	1–3	0·03–0·1
Winter maximum, varying from year to year	10–23	0·33–0·74
Nitrogen as inorganic nitrogen compounds (ammonia, nitrite and nitrate)		
Antarctic	*c.* 500	36
North Atlantic:		
Upper layers	< 10	0·7
Deep water	220	16
English Channel:		
Winter	*c.* 100	7
Upper layers in summer	< 10	0·7
Iron		
Total, almost wholly present as particles of ferric hydroxide	1–60	0·02–1·1
In solution at pH 8·3	4×10^{-7}	7×10^{-9}
Manganese		
Total, mostly as particles of oxides insoluble in very dilute acid	1–10	0·02–0·2
Soluble at pH 4·6	0·7–2·6	0·013–0·047

Iron, manganese and many other microelements are found heavily concentrated on or in plants and animals (pp. 142–7). In consequence they are liable to be depleted in the upper layers of the ocean, either to be returned after decomposition of the organisms at lower levels or to accumulate in bottom deposits. Such metals as titanium, manganese, nickel and cobalt, present at very great dilution in the water, are found in significant quantities in ocean deposits at great depths. The quantities of microelements, other than nitrogen and phosphorus, present in plankton organisms far from land have not yet been investigated.

PARTICLES IN SUSPENSION

In addition to living organisms, the sea contains a material quantity of both inorganic and organic particles in suspension.

Determination of the weight of particulate matter in suspension have been made in clear blue-green coastal waters off Plymouth, where the depth is 70 m., which are found to contain 0·4 to 2 mg. per litre; blue oceanic Atlantic water some hundred miles beyond the continental shelf contained 0·2–1 mg. per litre. About half this particulate matter was inorganic ferruginous clay-like material.

The turbid inshore waters, such as the southern North Sea, contain material quantities of organic and inorganic phosphate in the suspended particles.

It is probable that the inorganic particles play an indirect part in the changes taking place in the sea, due to the growth of bacteria attached to them (p. 67).

CHAPTER II

MIXING AND LATERAL TRANSPORT

CHANGES in salt content of ocean waters, with position, depth and time, are brought about in the main by greater evaporation taking place in subtropical latitudes and greater precipitation in polar seas. Wind, temperature differences, evaporation and precipitation give rise to currents, causing horizontal transport, most marked and rapid in the upper few hundred metres. Wave motion, convection currents, and turbulence set up by currents passing over an uneven bottom, give rise to vertical mixing. Considerable quantities of water upwell to take the place of water which is passing away as a current in some areas. It is evaporation, precipitation and movement, both horizontal and vertical, which regulate the distribution of salinity; simple diffusion of the salts in solution is extremely slow.

In order to present a picture of the constitution of sea water and the changes taking place in it in nature, some consideration must be given to the transport of water in ocean currents and the forces which either enhance or restrain mixing. This is equally necessary in any attempt to present the converse picture of how variations in the water affect the organisms and affect, even control, the fertility of the seas.

It is convenient to take the Atlantic Ocean as an example. The salt content of the waters is indicated in figs. 1 and 4. In high latitudes the salinity of the upper layers is frequently less than the salinity of the water below. The run-off from melting ice or the land fans out on the surface and sets up a current which is deflected by the earth's rotation and also grows deeper as it picks up water from below. In lower latitudes the salinity is usually greater in the upper layers; here evaporation has concentrated the water at the surface.

A complex system of currents is set up in the upper layers caused by wind, temperature differences, evaporation and precipitation (fig. 2). Friction and the force due to the earth's rotation influence their direction.

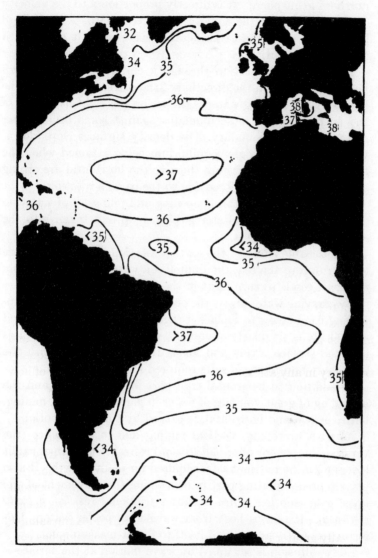

Fig. 1. Salinity (S ‰) in grams per kilo, of the surface
water of the Atlantic Ocean.

This force tends to deflect the current to the right in the northern hemisphere. It is directly proportional to the velocity of the current. At the equator the force is zero, while in high latitudes it may attain considerable magnitude; it varies with the sine of the latitude.

A vertical cross-section through a current in the northern hemisphere, seen in the direction of the current, shows that the light forward-moving water is deeper on the right-hand side of the current, where it accumulates against some other water mass or natural boundary. The density surfaces slope downwards towards the right, equilibrium being attained when the forces which tend to restore them to the horizontal are strong enough to balance the force due to the earth's rotation.

Hand in hand with this deepening and piling up of water on one side of a current, water is picked up and taken into the current as it progresses.

A noticeable effect of the earth's rotation is also seen where rivers run into the sea, the diluted sea water moving out from the coast tends to turn right in the northern hemisphere.

By carrying water away, the ocean currents lead to water upwelling from below to replace it (figs. 2 and 4). Both plant and animal life is singularly abundant, where upwelling penetrates into the surface layers and keeps them replenished with the phosphate and nitrogen salts required for plant growth.

In addition to horizontal transport in the currents and the upwelling of great volumes of water in particular areas, mixing, both vertical and horizontal, is caused by turbulent motion.

As with upwelling, vertical mixing due to turbulence (the vertical component of *eddy diffusion*) refreshes the upper sunlit layers with the nutrient salts required for plant growth. It also plays a predominating part in restraining the movement of one layer over another, since it causes frictional resistance (*eddy viscosity*). It carries heat from warmer to cooler water. Unfortunately it does not lend itself to direct measurement.

Eddy diffusion is set up (i) by wave motion at the surface—the orbital motion falls off with increasing depth, the decrease being less rapid in the open oceans with waves of greater distance from crest to crest; (ii) by convection currents set up by cooling and evaporation at the surface; (iii) by currents

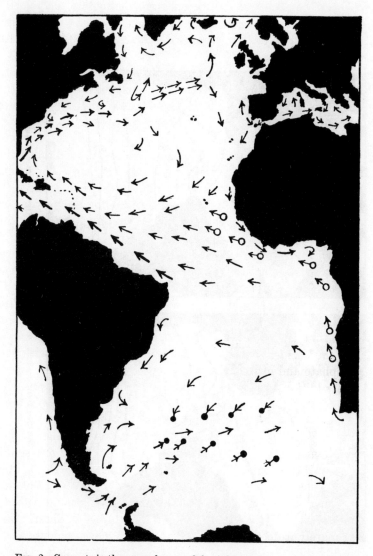

Fig. 2. Currents in the upper layers of the Atlantic Ocean. Arrows arising from a circle indicate water upwelling from below; arrows leading to a dot indicate water sinking below the upper layers.

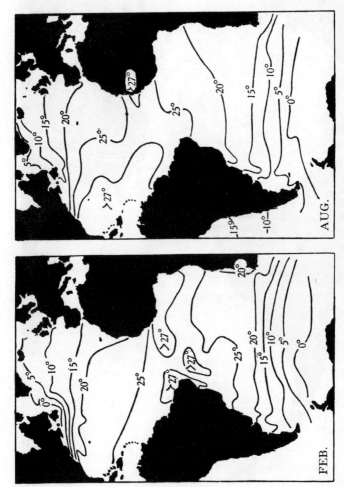

Fig. 3. Surface temperature, in degrees Centigrade, of the Atlantic Ocean during February and August.

passing over an uneven bottom, the turbulence so caused decreasing with increasing distance above the bottom. Turbulence developed in this way by to-and-fro tidal currents in relatively shallow seas often keeps the waters mixed vertically throughout

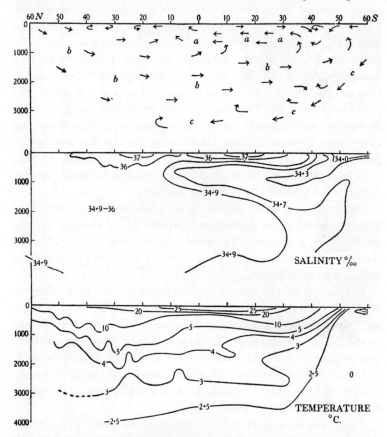

FIG. 4. Sections from 60° N. to 60° S. along the 30° W. meridian of longitude, showing the general trend of the currents, the salinity in grams per kilo and the temperature in degrees Centigrade. *a, a, a,* the 'Antarctic Intermediate Current'; *b, b, b,* the 'Atlantic Deep Water'; *c, c, c,* the 'Antarctic Deep Water'. Depth in metres.

the year. It is particularly noticeable where the current meets a hill or passes over a cliff in the bottom; then turbulence may be sufficiently great to cause ripples at the surface on a calm day some 100–150 m. above the bottom.

A secondary result of turbulence set up at the bottom is, by increasing frictional resistance in the water above, to cause a superficial current to change its course. An area of the sea where turbulence is relatively great acts as a natural boundary to an ocean current. Such areas are found where the deep ocean floor rises to the submarine plateau which surrounds the continents and are also found over submarine ridges. Such a ridge runs down the centre of the Atlantic; the path of the weakened westerly drift of the Gulf Stream is undoubtedly influenced by this, and the main flow towards Europe passes over it where the depths are greatest. The main current or drift of water into the Norwegian and Barents Sea likewise passes over the submarine ridge lying between Scotland and Iceland where the depths are greatest—that is, through the Faroe-Shetland Channel. In this way the contour of the sea floor has a marked influence on the direction of currents in water far above it.

Yet another type of motion takes place within the sea. The level, at which some particular temperature or salinity occurs, oscillates up and down. The range may exceed 200 m. at mid-depths in the open oceans. These *internal waves* will presumably enhance eddy diffusion.

The waters of the oceans are never still. It is only in some adjacent seas, cut off from the circulation of the oceans by a submarine ridge, that stagnant deep water is found. Such occurs in the Black Sea and some Norwegian fiords, where, as a result, the deep water becomes deoxygenated and sulphides are formed.

It remains to consider the conditions which damp down or hinder vertical eddy diffusion, that is, conditions which lead to *stability of the water column.*

Where the density of the water increases rapidly with depth, due to increasing salinity, decreasing temperature, or both, considerable work is required to mix the lighter water with the heavier water below. The vertical component of eddy motion is restrained and vertical eddy diffusion hindered. In low latitudes insolation warms the upper layers, and below a depth of some 100 m. there is usually a rapid decrease in temperature. Over the Sargasso Sea, the warm water, forced in from the encircling Gulf Stream system of currents, extends to greater depths. A layer of rapidly increasing density is known as the *thermocline* or *discontinuity layer*. In temperate and often in high latitudes, a thermocline is set up during the summer months, unless turbulence is so great that it never gets a chance of becoming established. This may happen in shallow seas with strong tidal currents, or in open oceans where a succession of gales follow each other. Over great tracts of the South Atlantic in the area of the Roaring Forties no thermocline develops in summer, whereas at an equal distance from the equator in the northern hemisphere a well-marked thermocline is set up.

It was at one time thought that the water above a well-marked thermocline was almost entirely cut off from the water below in respect to upward diffusion of dissolved salts, but some recent observations may lead to a modification of this view. However, only a sparse population of plant life is encountered above a marked thermocline which has persisted for any length of time, for the plants rapidly utilize the nutrient salts in the water and are left with an insufficient supply to maintain rapid growth.

The presence of a thermocline may have other far-reaching effects. Owing to reduced eddy viscosity within the thermocline the water above slides more readily over the water beneath, and horizontal transport is favoured. As a result of such movement, a change in salinity is often found at the thermocline.

It is interesting to follow the establishment and disappearance of a thermocline in temperate areas. In the semi-enclosed waters at the mouth of the English Channel, where the distance from crest to crest of the waves is relatively short compared with the open ocean, a strong well-marked thermocline develops at 13–14 m. in May or June when the weather is reasonably calm. A summer gale sends it down to some 18 m. or more, and subsequent hot calm weather may allow a second thermocline to develop above the first. The level at which the thermocline persists varies from year to year around an average depth of some 15 or 16 m. Its level, or depth below the surface, oscillates up and down some 2 m. due to internal waves. Towards the end of August the surface starts to lose more heat than it gains from insolation; this loss is due for the most part to evaporation. The upper warm layer cools gradually and the first autumn gale breaks down the thermocline. Then there is rather complete mixing of the water from top to bottom. The sequence of events is shown in fig. 5.

Farther seaward, in water having a depth of 2000–3000 m., the data available indicate that the thermocline is formed in summer at a deeper level, at some 25–30 m. Here the waves are longer from crest to crest, and the orbital motion which they set up will extend deeper. Fig. 6 shows the distribution of tem-

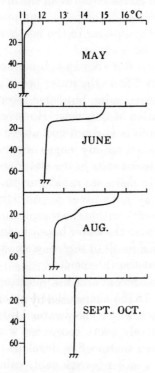

FIG. 5. The establishment and disappearance of a thermocline at a position in the English Channel. Depth 70 m.

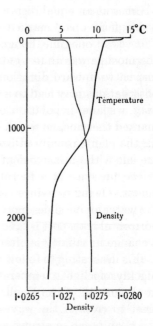

FIG. 6. Change in temperature and in density with depth at a position in the Atlantic, 360 miles south-west of the English Channel. Depth in metres.

perature with depth in June at a position some 350 miles south-west of the entrance to the English Channel. It may be inferred from it that when the thermocline breaks in autumn, mixing takes place to a considerable depth at this position in the open ocean.

These considerations indicate mechanisms by which the waters in the oceans are mixed with the water above and below. As a mass of water proceeds in a current, mixing also takes place with the water on either side; yet such a water mass may retain, for some years, characteristics such as having a salinity and temperature higher than that of the surrounding water—completion of horizontal mixing is very slow.

It is of interest to consider some differences between the open ocean and shallow inshore waters.

On approaching the coast of the continents from the deep water of the open oceans, the depth usually shoals rather rapidly from over 2000 m. to some 200 m. This *continental slope* is for the most part from 10 to 150 miles in width, except in the Arctic and between Europe and Greenland, where there are large tracts with depths intermediate between these values. The *continental shelf*, with depths of less than 200 m., may extend to a width of many miles and covers more than 6 % of the entire area of oceans. On passing the continental slope and proceeding over the shelf, the clear blue of the deep ocean changes to a more green hue, particularly in temperate and high latitudes. The amount of minute living organisms and particles of organic debris in the water becomes greater. The fauna living on and in the bottom plays a larger part in the changes taking place in the whole water column as the depths decrease over the continental shelf. Debris from dead organisms breaks down relatively quickly on reaching the bottom, and nutrient salts used by plants are regenerated nearer the surface than in deep-water areas. The bottom frequently contains a rich fauna, some 100 g. of living tissue composed of worms, protozoa and molluscs being no unusual quantity on and in a square metre of mud bottom. On nearing the coast, particularly where the sea is shallow and rivers enter the sea, the water is diluted and holds increasing quantities of sediment in suspension. The water is discoloured, its transparency reduced and the layer or zone in which there is sufficient light for plants to grow becomes thinner. Such 'inshore' conditions may extend many miles seaward over shallow depths, as in the southern North Sea, where tidal currents cause much vertical mixing and keep much sediment in suspension. During summer the whole column of water warms and there are no cool depths such as are found farther out to sea (fig. 7). During winter there is a tendency for the diluted water to remain banked up near the coast, whereas it runs out as a surface layer during summer. Then strong offshore winds may blow the warm water seaward, cooler subsurface water from offshore replacing it. Thus the inshore animals are subjected to a considerable annual range of salinity and temperature, with

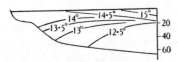

FIG. 7. Section, extending 20 miles south-west from Plymouth, showing the distribution of temperature (7 August 1924). Depth in metres.

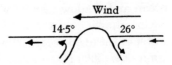

FIG. 8. Section showing how currents due to offshore winds affect the temperature of the surface water to leeward of the Galapagos Islands.

rather rapid changes. It is of interest that the distribution of many marine animals is limited by the range of temperature and has been linked with the salinity of the water, and it is remarkable that some species of fish are able to perceive very small differences not only in temperature but also in salinity, as little as 0·06 g. per kilo in water containing some 35 g. per kilo.

On passing up an estuary, a more or less rapid change in salinity is encountered. Since the dissolved salts in river waters are different in composition from those in sea water, the composition of the dissolved salts very gradually changes. In some

FIG. 9. Counter-current set up in estuaries where less saline water flows seaward in the upper layers.

estuaries, particularly where the bottom is uneven and there is a strong tidal flow, the water is kept well mixed from top to bottom. In others the fresher water flows out over more saline water, picking up water from below and carrying this out with it. This leads to a counter-current being set up near the bottom (fig. 9). In these estuaries where there is an increase in salinity with increasing depth, and where the whole mass of water moves up and down the estuary with the tide, there is a residual upstream movement of the bottom water due to this mechanism. Conditions in the River Tees estuary, which is of this type, have been very fully investigated. The diagrams in fig. 10, made from a great number of observations extending over several years, show the distribution of salinity at low and high water during summer and in winter, when the flow of fresh water is greater.

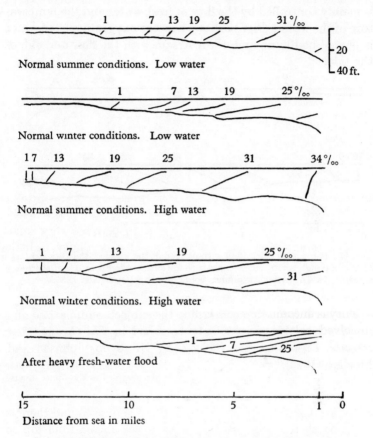

Fig. 10. Distribution of salinity in the estuary of the River Tees.
(After Alexander, Southgate and Bassindale, 1935.)

Measurements with floats and meters showed that the ebb tide was strongest in the surface layers and that the water ran out at the surface for a longer period than it flooded in, whereas the flood tide was strongest in the deeper layers where the water ran in for a longer period than it ebbed out.

A large quantity of organic matter is discharged into this estuary at various points between 5 and 11 miles from the sea, mainly at low water. In consequence of the rapid bacterial breakdown of this organic matter, the oxygen content of the water is reduced. The degree to which this reduction takes place is mainly controlled by the flow of fresh water and the temperature of the water. Fig. 11 shows how the oxygen-depleted water is moved bodily up and down stream with the flow and ebb of the tide.

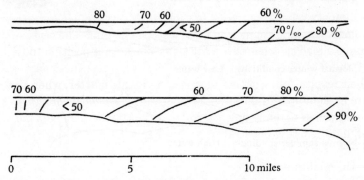

FIG. 11. Percentage saturation with oxygen of the water of the Tees Estuary, at low and high water under summer conditions, the temperature of the water being 13–16° C. (After Alexander, Southgate and Bassindale, 1935.)

Further information concerning the subjects summarized and simplified in this chapter may be found in *The Oceans, their Physics, Chemistry and Biology*, by Sverdrup, Johnson and Fleming (1942).

CHAPTER III

CHANGES IN COMPOSITION DUE TO PLANTS AND ANIMALS

THE PLANT AND ANIMAL COMMUNITY

THE plant life beyond the narrow fringe where sea weeds grow is almost entirely composed of unicellular organisms, varying in volume between about 1/50 and 1/50,000,000 mm³ The vegetable tissue which they contain is rather similar in composition to that of animal tissue, being rich in protein. Photosynthesis exceeds their respiration, with net increase in organic matter, when they are in the upper illuminated layer (the *photosynthetic or euphotic zone*) which extends downwards to a depth where some 3 g.cal. of light energy penetrates per square centimetre during the course of 24 hours. Below this depth—the compensation point —the plants continuously lose more substance by respiration than they gain by photosynthesis, and would eventually die. The depth of the photosynthetic zone varies greatly with transparency of the water, with latitude and season. In clear blue water of the tropics it extends to 100 m. or more, in turbid estuaries to a metre or less.

Occasionally some empty frustules of diatoms are found in the water where a natural mortality has occurred, but the fate of almost all these plants is to be eaten. One exception is the colonial alga *Phaeocystis*, which is not eaten, and so is enabled to build up a considerable standing crop in some areas from time to time, finally sinking and disintegrating.

It is. only in exceptional and very limited areas that whole plants sink uneaten and accumulate as a deposit on the bottom; in other, yet limited, areas there are deposits of broken and only partly digested diatoms, both in the deep water of the southern ocean and in some shallow areas.

On the floor of the deep ocean, and on the continental shelves, there is little accumulation of organic matter; either the remains of plants and animals have disintegrated before reaching the bottom, or having reached it, are soon collected and digested, mainly by filter-feeding, bottom-living animals. Over the continental shelf the to-and-fro tidal streams delay deposition by keeping fragments in suspension.

In the deep oceans few plants are found below 200 m. They have been eaten by the herbivorous zooplankton, many of which migrate upwards during the night into the photosynthetic zone and graze upon the growing plants. In their turn these herbivores are eaten by omnivorous and carnivorous species of zooplankton, and by pelagic fish.

If phytoplankton is plentiful, copepod zooplankton are notably voracious and void much undigested food. I have watched a *Calanus*, feeding in a suspension of diatoms, void faecal pellets at 20-minute intervals, and am told that this rate can be exceeded. Many faecal pellets have been found in the sea at the end of the spring outburst of diatoms; the number found below a square metre varied with the population density of the larger diatoms (Harvey, Cooper, Lebour and Russell, 1935). If phytoplankton is sparse, some species appear to feed on once-eaten fragments which have remained in suspension in the water (Mare, 1940). Apparently some of these fragments escape complete bacterial decomposition and reach the bottom even at very great depths, since bottom-living animals which presumably get their nourishment from organic detritus have been found there. At these great depths where the temperature is less than $2 \cdot 5°$ C., the food requirement of these animals must be very small, since they contain very little tissue, being mostly water and skeleton.

Over the continental shelf in areas of decreasing depth, the bottom-living fauna play an increasing part in the series of events by which plant tissue is finally converted into CO_2 and mineral salts. Many of these animals obtain their food by creating a current of water from which plankton and organic detritus is filtered. Others utilize organic detritus which has been deposited, and also protozoa and bacteria in the soil. Yet others are scavengers, living on pieces or maimed animals falling from above. In general this bottom fauna appears capable of dealing with much of the rain of organic detritus rather quickly, while bacteria and protozoa decompose the remainder.

Compared with the zooplankton community these bottom-living animals are large, respire at a lesser rate and grow slowly. The quantity of food which they are obliged to digest per unit dry weight of tissue to balance their losses by respiration is much less. Consequently the same quantity of food can maintain a greater bio-mass.

In areas where the sea floor is suitable for settlement of the larvae of filter-feeding animals (as lamellibranchs and many species of polychaetes), the nature of the sea floor permits a community of slow-growing, slow-respiring animals capable of supporting considerable populations of demersal fish.

Thus the nature of the sea floor below the continental shelf affects both the kind and quantity of animals maintained below unit area.

In tropic waters the round of events may proceed throughout the year at a fairly even rate, but in temperate seas there is marked seasonal variation in plant production and in the zooplankton population grazing it. This seasonal change follows a fairly regular pattern in most temperate areas. Early in the year, with increasing length of day, the plant population increases, reaching a maximum in spring. Meanwhile the zooplankton herbivores start to increase, until with rising temperature and ample food large broods develop which quickly graze down the spring maximum of phytoplankton. From then on during the summer months, heavy grazing and reduced growth rate due to insufficiency of available nitrogen and phosphorus, keeps the standing crop of plants at a low level. In late summer many of the predatory zooplankton living on the herbivores die out, permitting good survival of the final brood of herbivores in early autumn at a time when vertical mixing of the water is increasing, and bringing more available nitrogen and phosphorus into the photosynthetic zone, so permitting increased plant growth.

In high latitudes there is a single peak in both phytoplankton and zooplankton populations during the short summer.

Fig. 12, founded on observations in the English Channel, depicts the seasonal changes in the standing crop of phytoplankton and the biomass of zooplankton typical of a temperate sea. The organic matter present in the standing crop of plants

is roughly calculated from its chlorophyll content in a year when
no notable growths of (inedible) *Phaeocystis* arose. Such calculated values are crude since the proportion of chlorophyll to
organic matter in the plants is variable, and since the quantity
found in the water included chlorophyll in once-eaten plant
★ remains (Gillbricht, 1952). If exactly the same body of water
could be sampled at frequent intervals, marked and considerable
fluctuations would be found, since day-to-day changes in light

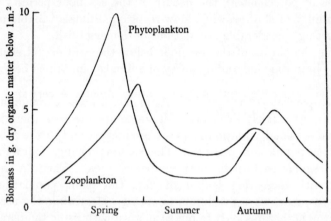

FIG. 12. Change in biomass of phytoplankton and zooplankton below unit
area, throughout the seasons, derived from observations in the English
Channel, 4 miles offshore.

intensity and in turbulence of the water affect the proliferation
of the plants, and since the average zooplankton population in
a small restricted area changes as brood follows brood and the
incidence of predators fluctuates. The distribution of zooplankton is patchy.

Ocean water contains a very small number of bacteria free in
suspension. To make good their high rate of respiration and
remain alive each individual needs to absorb a considerable
quantity of dissolved food daily. The concentration of dissolved
organic matter in ocean water is very low (3–6 mg. per litre),
and only about half of this is available. The standing population
of bacteria is very small, owing to this low concentration of food,
and probably to many being eaten by other small organisms.
In addition to these bacteria which are freely suspended in the

water, there are others living attached to the mucus covering of
phytoplankton and animals, and on particles of organic detritus.
Also any particles of inorganic detritus, of which there are many
even far out to sea, adsorb dissolved organic matter on to their
surfaces and so provide localized concentrations of nourishment
which permit bacteria to proliferate (p. 67).

ELEMENTS ABSTRACTED BY PLANTS

The composition of the plants shows the elements abstracted
from the water, and provides the first step towards showing the
actual compounds which are abstracted. In course of time, after
plants have passed through animals or have been converted into
animal or bacterial tissue, these same compounds are returned
to the sea water. This return is not quite complete. A very small
fraction may end up as bottom deposits.

TABLE 4. *Analyses of plankton diatoms*

C	N	P	
100	18·2	1·36	Bay of Fundy (Redfield, 1934)
100	15·6	2·26	Nova Scotia (Redfield, 1934)
100	14·2	1·4	Gulf of Maine (Waksman, Stokes and Butler, 1937)
	16	1·92	English Channel (Cooper, 1937)
100	15·8	—	North Atlantic (Von Brand, 1938)
Mean 100	16	1·7	
100	17	2·4	Mean value of various analyses of phytoplankton by Vinogradov (quoted by Fleming, 1940)
100	16	2·4	Mean value of various analyses of zooplankton

Analyses have been made of diatoms collected in temperate
seas during the spring, when available nitrogen and phosphorus
are ample for rapid growth. The most recent are shown in
table 4.

The relative proportion of these elements in the same species
is not constant unless growing under similar conditions. The
nitrogen and phosphorus content is dependent upon the concentrations of these elements in available form in the water.

It is noteworthy that the elementary composition of the plants is very similar to that of zooplankton animals.

It is also noteworthy that the ratio of nitrogen to phosphorus in the naturally growing plants which have been analysed is similar to the ratio of these elements in available form in the water in which they were growing (Harvey, 1927; Redfield, 1934).

Analyses have been made of seven species of unicellular algae, including one marine species, grown in culture. The conditions for growth were such that the supply of carbon dioxide in the water and the supply of available nitrogen and phosphorus were greater than in the sea. Nevertheless, on analysis (Ketchum and Redfield, 1949) all six species yielded remarkably similar values indicating a composition of

Protein   41–53 %
Carbohydrate 26–38 %
Lipid   20–27 %

In addition to abstracting carbon dioxide, phosphate and combined nitrogen from the water, diatoms in some areas extract almost all the dissolved silicate from the water (p. 147), while coccolithophores with calcareous exoskeletons surely play some part in reducing the 'carbonate alkalinity' of the upper layers in tropic seas (p. 161).

Analyses of diatoms collected in the English Channel and elsewhere show the presence of a material quantity of iron, probably micelles and particles of hydroxide held on their surfaces (p. 144).

Analyses of diatoms in Puget Sound show the presence of material quantities of manganese (p. 146). There is a rain of particles of manganese oxide deposited on the ocean floor, derived either from volcanic dust or as a result of adsorption on the surfaces of organisms. Deep-sea deposits also contain nickel, cobalt and titanium oxide, which may also in part be derived from organisms which have collected them.

Several trace elements in addition to iron and manganese are known to be essential for the growth of plants. These include copper, zinc, molybdenum and cobalt (possibly gallium). There is indirect evidence that copper is absorbed in material quantity
★ by phytoplankton, since its concentration in the English Channel

undergoes a seasonal variation (p. 141). The considerable quantities found in animals having haemocyanin as respiratory pigment are presumably derived from phytoplankton in the first instance.

There are no analyses showing that other microconstituents of sea water are collected in material quantity by phytoplankton, but analyses of seaweeds show notable quantities of Zn, Mo, Co, Ni, Pb, V, Ti and Cr (Black and Mitchell, 1952). There is no more than presumptive evidence that these elements may be carried downwards in the oceans due to adsorption on or absorption into organisms; their distribution with depth has not been investigated.

Some phytoplankton species, notable centric diatoms, contain much cell sap. These organisms have the same density as the surrounding water, thereby remaining in suspension while actively growing, the sap being thought to contain less calcium, magnesium and sulphate than the water (p. 102). The sap of the large unicellular marine algae, *Valonia* and *Halicystis*, have been analysed; the sap of *Valonia* is singularly rich in potassium, while the sap of *Halicystis* is rich in sodium, and contains less potassium, calcium, magnesium and sulphate than sea water. Plants can play no significant part in the redistribution of these major constituents of sea water, with the exception of Coccolithophores, which absorb and deposit calcium.

ELEMENTS ABSTRACTED BY ANIMALS

A quantity of calcium is deposited as shell around the coasts, and this shell contains a small but material quantity of phosphorus. In tropic areas much calcium is abstracted to form the skeletons of coral. Extensive deposits of pteropod and globigerina ooze occur down to depths of some 3000 m., below which deep-sea deposits contain less and less $CaCO_3$, which is more soluble in sea water under great pressure (p. 161).

Silicate is abstracted by some species of sponges and deposited as silica spicules. Radiolarian ooze occurs in tropical Pacific and Indian Ocean areas.

The body fluids of many species of Crustacea contain copper as haemocyanin, and the body fluids of *Tunicata* species are rich in vanadium, which may have been either abstracted by the animals or accumulated from food (p. 138).

Several animals have been found to be rich in one or more elements which occur at very great dilution in sea water. Thus *Cyanea* has been found containing 1·5 g. of zinc per kilo dry weight, and oysters are found which are rich in copper and coloured bright green as a result (p. 139).

In many species of invertebrates the composition and concentration of the salts in the body fluids are almost the same as in the surrounding sea water, although in the more highly developed cephalopods there is a greater proportion of potassium and in crabs and lobsters less magnesium.

The Carbon-Oxygen Cycle

Phytoplankton growing in the photosynthetic zone discharges into the water a volume of oxygen similar to the volume of CO_2 assimilated. Experiments by Sargent and Hindman (1943) indicate a ratio of CO_2 to O_2 which is close to unity, while Barker (1935) found a ratio of $0·95 \pm 0·03$.

The value of this ratio is of interest, since estimations of the production of plant organic matter in the sea have been made using the quantity of oxygen evolved by the plants. It is obvious that the ratio will be less if the plants are using nitrate as a nitrogen source than if they are using ammonium-nitrogen, also that it will be less the more protein the plants contain.

The elementary composition of *Nitzschia*, grown in culture and analysed by Ketchum and Redfield (1949), indicates a composition approximating to $C_{18}H_{30}N_2O_8$. The synthesis from CO_2, NH_4 and H_2O,

$$18CO_2 + 2NH_3 + 12H_2O \rightarrow C_{18}H_{30}N_2O_8 + 20O_2,$$

indicates a ratio of CO_2 to O_2 of 0·9. With nitrate as nitrogen source, the ratio becomes 0·84.

These cultured *Nitzschia* contained less nitrogen in proportion to carbon (13/100) than plankton diatoms caught in the spring, in consequence the ratios are both likely to be a little more than with naturally growing diatoms. There is a marked discrepancy between these calculated values and the observed values of the photosynthetic ratio.

As a result of plant growth the partial pressure of CO_2 in the water is lowered by a very small amount compared with the increase of partial pressure of dissolved oxygen.

Thus if $2 \cdot 24$ cm^3 of CO_2 is abstracted from a litre of ocean water, in equilibrium with the atmosphere at 16° C., its partial pressure falls by 12×10^{-5} atmosphere. If this quantity of oxygen is added, its partial pressure is increased by $45,000 \times 10^{-5}$ atmosphere.

In consequence the rate of interchange of CO_2 with the atmosphere is infinitely less than the interchange of oxygen. Hence a greater accumulation of CO_2 than loss of dissolved oxygen may be expected in a column of water.

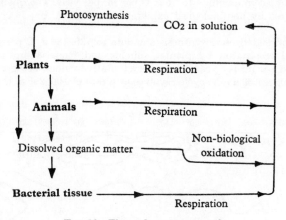

FIG. 13. The carbon-oxygen cycle.

Very little of the deposited organic detritus persists and is accumulated on the sea floor. On the continental shelf shell formation is slow, except in quite limited areas. In the top layer of the deep ocean deposits forming the floor of the Pacific, Arrhenius (1952) records 5–10 g. C per kilo, and calculates that it is accumulating at a rate of 2–20 mg. C per square metre annually.

Fig. 13 implies a state of equilibrium, but in nature production of plant tissue varies from day to day, and all the rates of change vary with the seasons. It is an overall picture of the average in a sufficiently large sea area over a sufficient length of time.

The term 'non-biological oxidation' is used in its widest sense, to include any oxidations which may be activated by enzymes excreted into the water as well as by other catalysts. Evidence has been found that enzymes exist free in salt water (p. 68); oxidative catalysts have been found in solution (p. 151).

It is interesting to speculate concerning the rate at which the cycle revolves in a sea area of moderate depth. In the English Channel, with an average depth of 70 m., the annual production of plant tissue (photosynthesis minus plant respiration) below a square metre is thought to contain around 100–150 g. C, and the average standing crop of phytoplankton per square metre to contain around 2 g. C.

The respiration of this plant tissue may amount to around 30 g. C per m.² per year (Riley, Stommel and Bumpus, 1949). An assessment has been made of the average quantities of animal and bacterial tissue below unit area, and of their respiration rates (Harvey, 1950). After modification due to subsequent observations, the computed quantity of CO_2 respired by them amounts to 75 g. C per m.² per year. The quantity of CO_2 set free by non-biological oxidation is unknown, but it is surely significant.

Thus estimated photosynthesis amounts to 130–180 g. C per m.² per year, and respiration by plants, animals and bacteria amounts to 105 + g. C per m.² per year. These very tentative values, being roughly similar, suggest that the cycle revolves at a rate of 100–200 g. C per m.² per year.

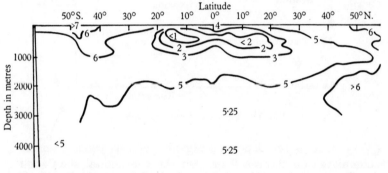

Fig. 14. Distribution of dissolved oxygen in cubic centimetres per litre in the eastern basin of the Atlantic between 50° N. and 50° S. (Data from *Meteor* Expedition, 1925–7.)

The distribution of oxygen in the oceans. The distribution of oxygen with depth in the eastern Atlantic is shown in figs. 14 and 15, which illustrate the oxygen-poor layer between 20° N. and 20° S. centred about a depth of 400–500 m. In the western Atlantic the layer is not so poor in oxygen but extends farther north below the Sargasso, at a greater depth. At the surface, dissolved oxygen is close to saturation with respect to the atmosphere. It follows the distribution of temperature, and shows a similar seasonal variation, there being some 8 cm.³ O_2 per litre in high latitudes and 4·5 cm.³ per litre in tropic surface waters.

Slight supersaturation is often found, due to photosynthesis. Deacon (1933) records 10 % supersaturation in the Antarctic, Sverdrup (1933) 12 % in the Arctic, and Cooper (1933) 8 % in the English Channel. In the Antarctic undersaturated water has also been found by Deacon, owing to oxygen-poor water welling

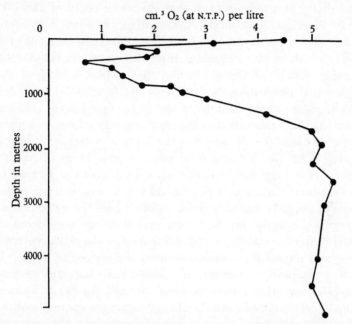

FIG. 15. Distribution of dissolved oxygen in cubic centimetres per litre with depth at 10° 12′ N., 26° 36′ W. (Data from *Meteor* Expedition, 1925–7.)

up from below. A detectable diurnal variation has been recorded by Jacobsen (1912). The reduced oxygen concentration in the oxygen-poor layer is due to respiration of animals, and of bacteria attached to falling particles of organic debris, exceeding the slow renewal by turbulence from the surface and from the water below. Seymour Sewell (1948) points out that zooplankton tend to congregate in this layer and may contribute materially to the processes. Seiwell & Seiwell (1938) have found that the sinking velocities of dead plankton organisms decreased to less than one-tenth after 14 days' storage, and that the oxygen-poor layers lie at a depth which is the most stable part of the water column. Redfield (1942) notes that oxygen-poor layers are found below relatively barren areas, in which the quantity of falling debris is not great, and that oxidation of this debris is likely to have been almost completed before falling so far. A study of the oxygen distribution in this layer has led to some interesting tentative conclusions. It is considered, from its salinity and temperature, that the water north of about 15 or 20° N. originates from the Arctic and moves slowly southward below the upper layers, gradually losing oxygen en route; the water south of this originates from the Antarctic (40–50° S.), moving slowly northward below the surface layers, as the Antarctic Intermediate Current. Seiwell (1934, 1937) has plotted the average oxygen content of the oxygen-poor layer between 1 and 32° N., and finds that the resulting curve drops to a minimum at about 15° N. and has the form of a regular series of undulations, the distance from peak to peak being about 200 miles. It is suggested that this periodicity arises as a result of the seasonal variation of phytoplankton growth, causing variations in oxygen content in those regions where the water masses were in the upper layers. These variations are considered to persist after the water has dipped down below the photosynthetic zone and moved great distances from the regions of origin. If this assumption is correct, it follows that the intermediate oxygen-poor layer moves at some 200 miles a year. Deacon (1933) has examined both the change in oxygen content and the change in salinity of the Antarctic Intermediate Current with change in latitude between 45° S. and 10° N., and obtained evidence of similar undulations.

A striking instance of the wealth of animal life in the oxygen-poor layer has been observed by Schmidt (1925) in the Gulf of Panama. Between 150 and 300 m. depth the water was only 10 to 2 % saturated with oxygen. On the other side of the Panama isthmus, in the Caribbean Sea, the oxygen-poor layer between the same depths was 60 to 50 % saturated, yet in the low oxygen-content water in the Gulf of Panama a much larger amount of animal life was found between these depths.

In the deep water below about 1500 m. there is a gradual decrease in oxygen between the Arctic and Antarctic. This deep water is considered to be derived from water which falls from the upper layers in high northern latitudes—in the Labrador

Sea and between Iceland and Greenland. In this deep water there is also found slightly more oxygen in the western than in the eastern basins.

In the eastern Pacific an oxygen-poor layer is found, centred around 300–400 m. depth, with exceptionally low oxygen concentration (Bruneau, Jerlov and Koczy, 1953). In such a layer the oxidation-reduction potential will be very different to that in the water above and below it (Cooper, 1937).

Riley (1951) has extended these concepts by quantitative assessments of the rate of oxidation of organic matter, of the current velocities and of the rate of vertical mixing at various depths in the Atlantic. He obtains assessed values which would account for the existing distribution of oxygen in this ocean. This mathematical analysis suggests that nine-tenths of the phytoplankton produced in the photosynthetic zone is oxidized to CO_2 in the upper 200 m., the remaining one-tenth being oxidized in mid and deep water. ★

Cooper (1933) has followed the seasonal change in dissolved oxygen in the English Channel; the concentration remains close to saturation with respect to the atmosphere, except that during the period of rising temperature and most photosynthesis between February and June the concentration near the surface showed slight supersaturation. Redfield (1948) has investigated the seasonal change in the Gulf of Maine and gives estimates of the exchange rate between sea and the atmosphere in summer and winter. This amounts to an accretion of about 300 litres of oxygen through a square metre of surface between October and late March. About 12 litres of this are attributed to the excess of photosynthesis over respiration, the remaining 288 litres being due to increased solubility as the temperature falls.

In some few exceptional areas in the open sea almost complete deoxygenation of the water takes place from time to time; off the west coast of Africa nutrient-rich water of the Benguela Current, upwelling to the surface, allows sudden intense growths of phytoplankton to take place at times. The decomposition of this vegetable tissue takes place more rapidly than oxygen enters from the air; organic matter is deposited, and sulphate-reducing bacteria thrive in the deposit under anaerobic conditions. The sulphide formed dissolves in the water above and persists at depths where the water remains deoxygenated. On occasions these events cause wholesale mortality of fish in the Walfish Bay area.

The oxygen content of the deep water of adjacent seas, cut off from the ocean circulation by submarine ridges, is of interest. In the Mediterranean the deep water is poorly oxygenated, while in the Black Sea the deep water is devoid of oxygen and hydrogen sulphide is present in quantity. Poor oxygenation is also found in the deep water of some fjords and in pockets in the Baltic.

The solubility and estimation of oxygen, and the rate of interchange with the atmosphere are summarized on p. 183.

The distribution of pH and of carbon dioxide in the oceans. The concentration of total carbon dioxide in solution in oceanic waters bears a rough inverse relation to their pH. On

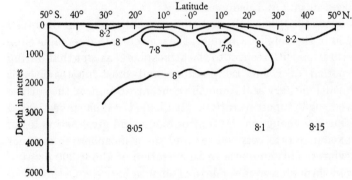

FIG. 16. Distribution of pH in the eastern Atlantic.
(Data from *Meteor* Expedition, 1925–7.)

comparing figs. 14 and 16 a general resemblance is seen between the distribution of dissolved oxygen and pH. It is also seen on comparing figs. 15 and 17; at this position there is some 5–6 cm³ per litre more carbon dioxide in the water of the oxygen-poor layer than at the surface.

In seas of moderate depth lying over the continental shelf, there is a seasonal change in the total CO_2 in the water. Observations through a year in the English Channel have shown a reduction of 2·7 cm³ CO_2 per litre between February and June.

(This is commensurate with photosynthesis having exceeded respiration below a square metre by 100 g. C in this period, assuming that the same water occupied the position during the five months, which it never does.)

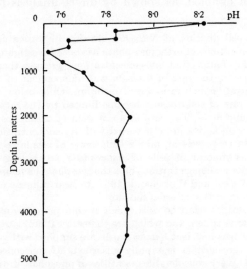

FIG. 17. Distribution of pH with depth at 10° 12′ N., 26° 36′ W. (Data from *Meteor* Expedition, 1925–7.)

THE PHOSPHORUS CYCLE

Uptake of phosphorus by plants. Although most of the phosphorus is absorbed by phytoplankton as orthophosphate ions, there is reason to believe that some may be absorbed as molecules of dissolved organic phosphate.

Chu (1946) found that *Nitzschia closterium* (*Phaeodactylum*) in bacteria-free culture could utilize glycerophosphate and inositol hexaphosphate, a stable storage product present as phytin in many plants. They also utilized organic phosphorus compounds obtained by extracting actively growing *Laminaria* in hot water. They did not absorb pyrophosphate ions.

When *Nitzschia* during growth has used all available phosphate in the water, cell division and photosynthesis continues for several divisions, the cells become phosphorus-deficient (Ketchum, 1939). Direct confirmation of Chu's observations have been obtained by adding glycerophosphate and also inositol hexaphosphate to bacteria-free phosphorus-deficient *Nitzschia* and storing them for three hours in darkness. During this short period the phosphorus content of the cells doubled or more than doubled, as shown by direct analysis (Harvey, 1953a, b).

The rate of cell division of *Nitzschia* growing in waters containing different concentrations of orthophosphate has been investigated (p. 96). Ketchum (1939) found that at concentrations greater than 16 mg. phosphate-P per m³ the rate of division was independent of the phosphorus concentration in the external medium, while below this concentration the rate of cell division became limited by the rate the cells could absorb phosphate. The rate of absorption of orthophosphate by this diatom during active growth varied with its concentration in the water, provided there was an ample sufficiency of available nitrogen. The phosphorus content of cells of *Astereonella japonica* growing in waters with different concentrations of orthophosphate have been shown by Goldberg, Walker and Whisenand (1951) to bear a linear relation to the concentration in the external medium.

These observations relate to cells which are not phosphorus-deficient when phosphate is added, and which are thereafter illuminated.

When phosphorus-deficient *Nitzschia* cells are supplied with phosphate and stored in the dark, it is rapidly absorbed. If the concentration supplied is great, very considerable quantities of phosphate are absorbed and held in the cells in organic combination. Similar conclusions have been reached by Matsue (1949) using *Skeletonema costatum*, which absorbed more rapidly at 25° C. than at 10° C. If such heavily loaded cells are then illuminated, a long lag period has been observed before cell division starts, and during this lag period phosphate was excreted from *Nitzschia* cells into the water (Spencer, 1954). In experiments by Scott (1945) phosphorus-deficient *Chlorella* were supplied with phosphate and illuminated. During the first 27 hours' illumination the cellular phosphorus increased sevenfold, and thereafter there was a loss into the water before cell division started. He also observed that when phosphate was absorbed by phosphate-deficient *Chlorella* an equivalent quantity of potassium was also absorbed.

There is some experimental evidence that when *Nitzschia* is growing in water having a low phosphate concentration, the growing cells may be somewhat phosphorus-deficient. This diatom was grown in culture until the concentration of phosphate in the medium was reduced to 5 mg. phosphate-P per m³ Additions of phosphate were then made to portions of the culture which was thereafter stored for 16 hours in darkness. The resulting increase in cellular phosphorus is plotted against concentrations of phosphate in the external medium in fig. 18. At the conclusion of the experiment the cells were resuspended in phosphate-free sea water into which they discharged a negligible quantity of phosphate (F. A. Armstrong, private communication).

It appears from these observations that when *Nitzschia* and *Chlorella* are supplied with high concentrations of phosphate, much of the intake is converted and firmly held as an organic storage product (Matsue, 1949), and that quantities of this storage produced in excess of a threshold value are dephosphorylated and discharged as phosphate under the influence of light.

A fresh-water diatom has been found capable of absorbing phosphate from concentrations containing less than 1 mg. phosphate-P per m³ of lake water and using this store to make many divisions (Mackereth, 1953).

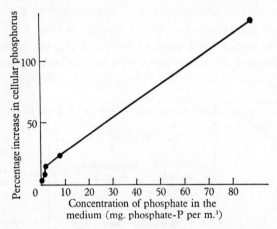

FIG. 18.   Relation of increase of cellular phosphate in *Nitzschia* to concentration of phosphate in medium. (Data from F. A. Armstrong.)

From these various observations on a few species of phytoplankton, it appears likely: (i) that they can absorb phosphate as quickly as they need it for rapid growth when its concentration in the water exceeds a threshold value whose magnitude lies in the region of 16 mg. phosphate-P per m³; (ii) that they can continue absorption of phosphate and its conversion into organic phosphorus compounds throughout both day and night; and (iii) that they can build up a reserve of storage product which cannot be used directly for further syntheses without prior dephosphorylation, and that light sets free or activates the phosphorylase concerned.

Discharge of phosphorus from plant tissue, etc. Much of the organic phosphorus in phytoplankton tissue is dephosphorylated by the phosphatases in the cell, yielding orthophosphate,

when the plants die (Matsue, 1949). In the growing frond of *Laminaria*, Chu found that only 1 % of the phosphorus in it was in the form of orthophosphate, while from a moribund frond 80 % of the phosphorus dissolved in hot water as ortho-
★ phosphate.

In nature nearly all the phytoplankton is eaten by animals, largely by zooplankton, and part is voided without being digested. Analyses of the faecal pellets of copepods indicate that most of the phosphorus does not remain in the voided particles of plants but dissolves in the sea water. Part of this persists in the sea for a time as dissolved organic phosphorus compounds, and part is doubtless dephosphorylated by vegetable phosphorylases while in the animals' guts or while in faecal

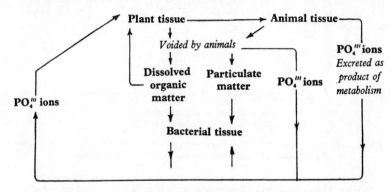

FIG. 19. The phosphorus cycle.

pellets. Some of the phytoplankton is digested and excreted by the animals as orthophosphate, excreted continuously whether the animals are well fed or starved.

In an experiment the diatom *Skeletonema costatum* was crushed in the presence of chloroform to reduce or stop bacterial decomposition. After storage for three hours, 35% of the phosphorus was in the form of inorganic phosphate, whereas, when heated to destroy phosphorylase before crushing, only 3% was in the form of inorganic phosphate. After crushing, some 75% of the cellular phosphorus was in water-soluble form.

The particles of phytoplankton tissue, also of animal tissue, which have been voided by animals, are either eaten again by other animals or broken down by bacteria.

During the summer when phytoplankton is sparse in the English Channel, copepods are seen with their guts full of brown detritus which originated as voided particles of plants and animals. Throughout the year these detritus particles form a large part of the food of filter-feeding animals living on or in the bottom.

When bacteria proliferate, as on particles of organic detritus, the bacterial tissue which is formed is exceptionally rich in ph os- phorus, and in order to remain alive the cells require a continued supply of organic carbon to make good their respiratory losses.

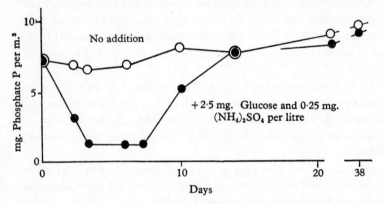

FIG. 20. Changes in phosphate concentration during storage in a sample of sea water, filtered through Whatman no. 3 paper. The lower curve shows the effect of increased bacterial growth due to the addition of phosphorus-free nutrient.

If the organic food contains insufficient phosphorus, they can utilize orthophosphate in solution in the water to build new bacterial tissue. When the supply of organic carbon fails, the cells die and rapidly set free their contained organic phosphorus as orthophosphate. The data shown in fig. 20 indicate how quickly this may occur.

The cycle of events causes an overall downward movement of phosphorus compounds in the sea. ★

The diagram of the phosphorus cycle represents one which is closed. This is not strictly true.

There is evidence that in shallow seas a temporary 'lock-up' of some phosphorus occurs during spring and early summer.

This may be due in part to deposition of detritus which takes some months to return all its contained phosphorus to the water as phosphate.

Below seas of moderate depth there is a deposit of shell accumulated over the ages, and this contains some 0·3 % of phosphorus. In the ocean floor of the Pacific at depths of several thousand metres, Arrhenius (1952) records 0·2–0·9 g. P per kilo of the top layer of deposit. He calculates that phosphate is accumulating in the deposits at an average rate of 0·5–4·0 mg. ★ P per m? annually.

Some interchange of phosphate between the offshore sea floor and the water above is considered by Cooper (1951) to occur in a sea area having a floor of very fine particles. Stephenson (1949) has studied the release of phosphate from the mud of a polluted estuary.

Turbid inshore waters, as in the southern North Sea, can contain material quantities of particulate phosphate which dissolves in dilute acid, as in the 0·28N acid commonly used in estimating phosphate (Kalle, 1937; Armstrong and Harvey, 1950).

Deposits of calcium phosphate containing fluoride have been found in the Pacific in warm shallow water off the California coast (Dietz, Emery and Shepard, 1942).

Deposition of phosphate is offset by inflow from rivers. In rivers of temperate latitudes which are not turbid, most of the phosphate is used by plants before the rivers reach the sea during the summer months, so most of the sea's enrichment takes place in winter. The influence of rivers on the phosphate content of the sea into which they flow is rarely obvious, dilution of the fresh water with great quantities of sea water being rapid.

When either raw or filtered sea water is stored in bottles, dissolved organic substances are adsorbed on the glass surface, notably at the meniscus. This causes local concentration of food for bacteria in which these proliferate very rapidly. In consequence, with some waters, the phosphate in solution is reduced—this happens when the bacterial food is poor in phosphorus, and in order to make new tissue the bacteria absorb phosphate from the water. The decrease of dissolved phosphate is much accentuated if a trace of sugar or asparagin is added to the water. When most waters are stored, phosphate is set free from organic combination, there being more than is requisite in the adsorbed organic substances for the amount of new bacterial tissue which can be formed and maintained by the adsorbed food. Although the tissue is exceptionally rich in phosphorus, much organic carbon is required daily to make good its daily loss as respired carbon dioxide. The increase in dissolved phosphate can be greatly accentuated by adding a trace of nucleic acid or other stable organic phosphorus to the water.

Changes in phosphate concentration of sea waters during storage can be greatly reduced, but not stopped, by saturating with chloroform and cold storage. Table 5 shows how they were still further reduced and almost entirely stopped by saturation with chloroform and heating to 56° C. in order to inactivate enzymes in a water to which nucleinate had been added, and which was stored for 24 days after treatment.

TABLE 5. *Changes in phosphate concentration (in mg. phosphate-P per m³) in filtered sea water enriched with a trace of sodium nucleinate and stored in glass bottles at 20° C.*

	Initially	After 11 days	After 24 days
Enriched water alone	4·2, 4·3, 4·2	26·3	—
Enriched water saturated with CHCl₃	4·2, 4·3	7·6*	—
Enriched water saturated with CHCl₃ and heated to 56° C.	4·9, 4·8	5·6, 5·4	5·2, 5·35

* A similar increase during storage was found in subsequent experiments.

In shallow seas where strong winds from time to time stir the bottom deposits, a significant quantity of phosphate passes into the water above. Interstitial water between sand particles has been found to contain considerable quantities of phosphate in solution (J. H. Oliver, private communication), which was probably set free on autolysis of the rich bacterial flora attached to sand grains. In addition, at depths within the deposits where the pH is less than about 7·2, phosphate is likely to be present as ferric phosphate which will hydrolyse and set free phosphate if brought into contact with the water above, having a pH of 8 or more. ★

There is another aspect of phosphorus metabolism by algae which deserves consideration, since it plays a part in experiments where labelled phosphate ^{32}P is used. Interpretation of the results is not necessarily straightforward, because interchange of phosphate between that in the external medium and that present in the cells, both inorganic and organic, almost certainly takes place. Since the cells are rich in ^{31}P they tend to 'collect by interchange' relatively large quantities of ^{32}P from an external medium containing only minute traces of labelled phosphate. By analogy with muscle fibres (Causey and Harris, 1951) much of the labelled phosphate which is 'collected by interchange' passes into stable organic combination.

Distribution of organic phosphorus in the sea. Most of the phosphorus is present as dissolved orthophosphate, except during summer in the upper layers, when most may be present as dissolved organic phosphorus compounds. In clear offshore sea water a quite small proportion of the total is present in the standing crop of plants, in zooplankton and in organic detritus in suspension. However, in turbid waters, such as are found near the coast of the southern North Sea, quite considerable quantities may be present in detritus (Kalle, 1937; Harvey, 1950).

A further proportion is present in fish and in bottom-living animals.

A sample of raw sea water collected for analysis of 'total phosphorus' includes that in plants, in organic detritus particles and in the smaller zooplankton organisms. In many sea areas the major part of the biomass of zooplankton consists of these smaller organisms. These, together with the plants, are collected in samples obtained with a closing water bottle. The 'total phosphorus' in the water samples is determined as phosphate after wet combustion (Redfield, Smith and Ketchum, 1937) or after acid hydrolysis at 140° C., which has been found to set free as phosphate about 97 % of the phosphorus contained in living tissue (Harvey, 1948).

An attempt has been made to assess the average biomass of fish and bottom-living animals below unit area in the western part of the English Channel (Harvey, 1950). Fig. 21 is based on these tentative assessments and on observations of the quantity and phosphorus content of phytoplankton, zooplankton and detritus, and of 'total phosphorus' in the water.

The diagram undoubtedly contains considerable errors due to the relatively small number of observations and the conjectural assessments, notably of fish, on which it is founded. It does not illustrate any seasonal increase in the bottom-living fauna during the spring months, or any deposition during the spring of detritus which takes some months to return all its contained phosphorus to the water as phosphates (see below). However, it does provide a rough picture of how the phosphorus is partitioned in these waters, having an average depth of 70 m. and during years when the integral mean concentration of 'total phosphorus' in the water column varied around 12·5 mg. P per m³

The water occupying a position in the Gulf of Maine from time to time during the course of a year has been analysed by Redfield, Smith and Ketchum (1937). Their data for winter and summer is shown in fig. 22. The quantity of phosphorus present in particulate matter—plants, zooplankton and detritus—was

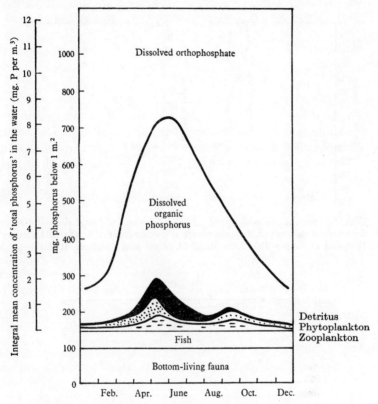

FIG. 21. Distribution of phosphorus in a water column 70 m. deep in the English Channel.

greatest in the upper 30 m., as much as 3 mg. P per m.³ in the summer and 1 mg. P per m.³ in winter, while the deeper water contained 1 mg. or less throughout the year.

At a station in the north-east Atlantic analyses of the water at 1000 m. depth and below showed insignificant differences between phosphate and 'total phosphorus'. The distribution with depth is shown in fig. 23.

The water occupying a position 20 miles offshore in the English Channel has been collected at intervals and analysed for phosphate and 'total phosphorus'. Typical distributions
★ with depth during winter and summer are shown in fig. 24.

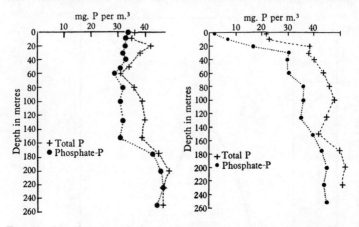

FIG. 22. Distribution with depth of phosphate and of 'total phosphorus' in the water of the Gulf of Maine in winter and in summer. (Data for 26 February 1936 and 21 August 1936 from Redfield, Smith and Ketchum, 1937.)

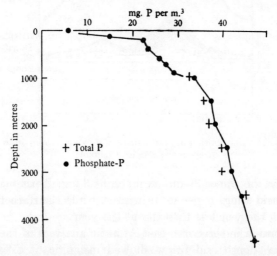

FIG. 23. Distribution with depth of phosphate and of 'total phosphorus' in the eastern North Atlantic, 46° 28′ N., 8° 06′ W., September 1952. (Data from F. A. J. Armstrong.)

Several features of interest arose during the course of this survey. (i) A device was used which rapidly sucked in water from around 5 cm. above the bottom. This water, in spite of its greater content of detritus, was not notably rich in 'total phosphorus'. There was no phosphorus-rich slurry over the bottom in this area. (ii) It has been observed that water collected at the sea surface is often notably rich in phosphate

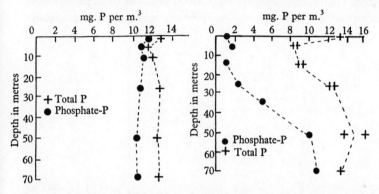

Fig. 24. Distribution with depth of phosphate and of 'total phosphorus' in the water of the English Channel, 20 miles offshore, in winter and summer. (Data for 5 January 1949 and 31 August 1948 from Armstrong and Harvey, 1950.)

and in total phosphorus, particularly during the winter. Suspended matter collected by filtration from surface water was found to be rich in phosphate; on two occasions it accounted for 5·5 and 2·2 mg. P per m³. The cause is not obvious; possibly dust from the atmosphere becomes trapped by surface tension at the air surface, or particles in suspension may attach themselves to minute bubbles of air during rough weather and these rising to the surface remain entrapped. Both Söderström (1924) and Wilson (1932) have observed that where larvae of *Polygordius* and of *Owenia* come into contact with the surface they are held there or even torn apart by the surface forces. (iii) During the years 1947–9 the water occupying the position at each date when samples were taken was not the same, as shown by changes in salinity. Yet, in spite of this, and in spite of much higher concentrations in water farther west and lower concentrations in waters up channel, the integral mean concentration of 'total phosphorus' in the water column did not fluctuate wildly (fig. 25).

However, the values for these years, and for subsequent years, indicate a rather regular decrease during the spring months amounting to 2–3 mg. P per m³. What has happened to this 140–210 mg. P which has disappeared from the water column? There is no evidence of a commensurate spring increase in the bottom fauna, nor evidence that water of lesser total phosphorus content enters the area regularly every spring and is gradually

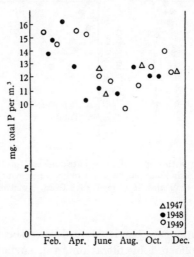

FIG. 25. Integral mean concentration of 'total phosphorus' in a water column, 70 m. deep, 20 miles offshore. (After Armstrong and Harvey, 1950.)

replaced later, but it is possible that the light phosphate-deficient upper layer has piled up before the prevalent south-westerly winds. Perhaps there is a deposit of phosphorus containing organic matter which requires several months for complete breakdown by animals and bacteria.

When the percentage of organic phosphorus in the 'total phosphorus' is plotted for these years, it is seen to follow a rather regular seasonal pattern, rising from c. 10 % in winter to 50 or 60 % in May. Thereafter there is a slow regular decrease, the rate of its breakdown to phosphate exceeding the rate at which it is being produced (fig. 26).

What proportion of this organic phosphorus is used as such by plants, and what proportion dephosphorylated by bacteria or hydrolysed without the agency of living organisms, is unknown.

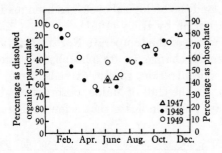

FIG. 26. Seasonal variation in the percentage composition of the phosphorus in a water column, 70 m. deep, 20 miles offshore. (After Armstrong and Harvey, 1950.)

Distribution of phosphate in the sea.

Evidence so far obtained indicates that the phosphorus in the oceans is almost entirely dissolved orthophosphate at 1000 m. depth and below. The temporary lock-up as organic phosphorus, and its downward movement due to particles sinking and to the vertical migration of zooplankton, is almost limited to the upper 1000 m. In this lies the zone of oxygen-poor water (p. 33) which is also rich in phosphate (Clowes, 1938; Deacon, 1933; Redfield, 1942; Riley, 1951). Below 1000 m. vertical eddy diffusion tends to equalize concentration between water layers of different origin.

Distribution in the ocean is the final state brought about by lateral currents, vertical eddy diffusion and the downward movement due to the cycle of biological events.

In the Atlantic, where the trend of the movement of the upper layers is from south to north, the lateral transport can be represented as in fig. 27. As a consequence less phosphate would be expected in the deep water of the Northern Hemisphere than in the Southern Ocean, where it upwells and enters the north-going current systems. The Atlantic Deep Water, on its slow passage south, gains phosphate from above and also from the north-going Antarctic Deep Water lying below it.

4

Owing to these lateral movements and to the downward movement caused by the biological cycle, the Southern Ocean is a reservoir of plant nutrients (see table 6).

An analysis of the parts played by lateral movement, vertical eddy diffusion, and the biological cycle has been made in an outstanding contribution by Riley (1951).

The surface water of the temperate North Atlantic undergoes a regular seasonal change in its phosphate content; during summer the concentration rarely exceeds 3 mg. P per m³ (Seiwell, 1935). In the Gulf of Maine, observations by Bigelow, Lillick and Sears (1940) indicate that a few milligrams remain.

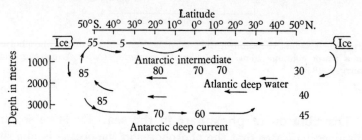

FIG. 27. Current systems and distribution of phosphate in mg. P per m³ in the Atlantic.

They found more complete exhaustion of nitrate and nitrite, while Redfield and Keys (1938) found complete exhaustion of ammonium-nitrogen in the surface layers during summer. These observations suggest a slight surplus of phosphates over available nitrogen for the plants' requirements in this area.

The distribution of phosphate in the Mediterranean is of interest. The water is replenished by surface-layer water from the Atlantic with a low phosphate content, which enters through the Straits of Gibraltar. Mediterranean water of higher salinity which has been concentrated by evaporation passes out through the Straits. This mechanism involves a continuous loss of phosphate. It is found that the concentrations of both phosphate and salts containing nitrogen are low in the deep water of the Mediterranean, compared with the deep water of the open oceans.

Bernard (1939) has followed the changes in phosphate in the upper layers off Monaco, and found that the concentration at the surface rarely exceeds 0·5 mg. P per m³, while even at a depth of 350 m. it does not rise above 2·5 mg. These upper layers were found to contain a relatively abundant supply of nitrate. The concentration of nitrogen-containing salts, relatively high compared with that of the phosphate, is attributed by Bernard (1939) to refreshment of the sea with river waters and rain.

TABLE 6. *Phosphate-phosphorus in mg. P per m³, in the water at positions along the 30° W. meridian of longitude during April–May* 1931

(From data published in *Discovery Reports*, vol. 4, after correction for salt error.)

Depth (m.)	57° 36′ S.	53° 33′ S.	46° 43′ S.	38° 10′ S.	21° 13′ S.	3° N.	9° N.	14° 27′ N.
0	62	—	47	5	0	0	0	—
20	64	—	47	7	0	0	0	—
40	63	60	47	7	0	0	0	—
60	74	60	48	7	2	0	0	—
80	76	67	52	7	3	15	14	2
100	77	70	55	15	4	42	33	12
150	83	77	60	18	5	46	44	37
200	85	77	65	18	6	50	45	40
400	83	77	67	45	31	69	52	59
800	83	77	81	58	64	77	80	78
1500	80	77	79	86	54	46	53	52
2500	83	77	73	78	42	40	45	45
3200	85	—	—	—	—	—	—	—
3500	—	80	78	82	41	—	46	46
4500	—	—	81	—	55	—	48	48
4900	—	—	—	—	69	—	—	—
5300	—	—	—	—	—	—	—	50

In the Adriatic, Ercegovic (1934) has found a seasonal variation in phosphate: the concentration in the surface water changing from less than 0·2 to 3 mg. P per m³ and at a depth of 90 m. from 1 to 3·2 mg.

The seasonal variation of phosphate at a position in the English Channel has been followed at monthly intervals for a number of years. From winter maximum values varying between 21·5 and 10 mg. P per m³ it has fallen in some years to less than

0·5 in the upper layers, in other years 2 or 3 mg. remain (Atkins, 1923–30; Cooper, 1933, 1938). The changes which took place during 1924 are shown in fig. 28. In the southern hemisphere, a similar seasonal variation has been found by Dakin and Colefax (1935).

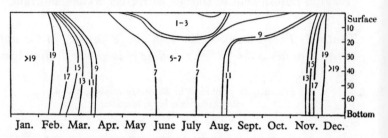

FIG. 28. Variation in phosphate during 1924 at a position in the English Channel. Numbers indicate milligrams of phosphate-phosphorus per cubic metre of water. Depth in metres.

The phosphorus cycle in lakes. The cycle of phosphorus in many lakes is quite different from that in the sea. Whole algae fall to the bottom, there being a much smaller zooplankton population to eat, crush or digest them while in suspension, and less bottom fauna to eat and digest them after they have settled. For some odd reason the algae are only very slowly decomposed by bacteria and protozoa, although both unicellular and filamentous algae are extremely rich in protein. Large quantities of phosphorus remain locked up in the algal deposit. Where anaerolic, ferrous ions appear in the deposit and when these diffuse out, together with phosphate, and penetrate into oxygenated water above, the ferrous iron is oxidized and ferric phosphate is precipitated. Since the water is rarely alkaline like sea water, the ferric phosphate is not hydrolysed.

The water of lakes and ponds is in consequence often rich in salts containing nitrogen, but very poor in phosphate. If they are fertilized with phosphate, this is rapidly consumed by algae and water weeds, and much of the added phosphate remains locked up for long periods.

THE NITROGEN CYCLE

The nitrogen cycle in the sea follows a course similar to that of phosphorus. The firstfruit of remineralization from organic nitrogen is ammonium ions, with which are included any undissociated NH_3 or NH_4OH, the relative proportions depending upon pH. This breakdown to ammonia is brought about by proteolytic enzymes, secreted into the digestive tracts of animals or excreted by bacteria. Mere crushing of the plant cells does

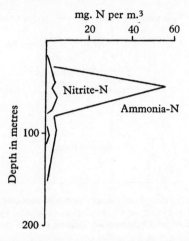

FIG. 29. Concentration of ammonium- and of nitrite-nitrogen in the Gulf of Maine. (After Redfield and Keys, 1938.)

not set free these enzymes as it does phosphorylase. For this reason alone, remineralization of combined nitrogen is likely to be slower than remineralization of phosphorus, and proportionately less nitrogen than phosphorus is likely to be remineralized within the photosynthetic zone. If ammonium does not happen to be absorbed by plants, it is oxidized with loss of heat to nitrite and then to nitrate, this oxidation being activated by bacteria. The oxidation to nitrite takes place immediately below the photosynthetic zone in the open oceans (fig. 29; see also Thompson and Gilson, 1937) and is followed by oxidation to nitrate (p. 77).

In deep oceans, and in shallower seas during winter, nearly all the combined inorganic nitrogen is present as nitrate.

Dissolved organic nitrogen has been found in considerable quantity in sea water (p. 150); its seasonal variation has not been investigated. Judging by the decrease in inorganic nitrogen compounds (fig. 32) a marked increase in dissolved organic nitrogen would be expected during spring and early summer.

The cycle is not completely closed. The sea receives a small annual addition of combined nitrogen from rivers and rain.

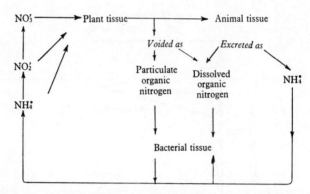

Fig. 30. The nitrogen cycle. For clarity, possible extracellular oxidations activated by bacterial enzymes or by oxidation catalysts, and also possible absorption of dissolved organic nitrogen by plants, are not shown.

Clark (1924) estimates that rain brings some 28 mg. of nitrate-nitrogen and 56–240 mg. of ammonium-nitrogen to each square metre of the sea surface annually. Deacon (1933) found 10 mg. of nitrate-nitrogen per m^3 of tropical rain, Braarud and Klem (1931) 24–34 mg. per m^3 in snow on the coast of Norway. Braadlie (1930) has determined the nitrate- and ammonium-nitrogen in rain and snow throughout the year on the Norwegian coast, finding an average of some 135 mg. NH_4-N and 75 mg.
★ nitrate-N per m^3.

The quantity of combined nitrogen discharged into the sea by rivers is much less during summer than winter. The annual addition to the sea from land drainage is considerable, but compared with the store of combined nitrogen in sea water this addition must be very small indeed. Riley (1937) quotes analyses showing Mississippi water to contain on the average throughout the year:

NH_4-N	20 mg. per m.³	Nitrite-N	5 mg. per m.³
Albuminoid-N	350 mg per m.³	Nitrate-N	200 mg. per m.³

Sea water, except in the upper layers where plant growth is taking place, contains about half this quantity of combined nitrogen.

Bacterial fixation of dissolved nitrogen by *Azotobacter* and *Clostridium* in the water, on plant organisms, and in bottom deposits, has not been demonstrated to take place in the oceans.

The loss of combined nitrogen from the cycle appears to be rather insignificant. There is slow deposition of marine humus, which is very resistant to attack by bacteria and persists for long periods. It is similar in composition and resistance to a ferri-ligno-protein complex.

There may be some small annual loss through the agency of denitrifying bacteria, but no evidence has been found which indicates that this takes place, with the possible exception of a few inshore positions where there is a deposition of organic matter on the bottom and a considerable concentration of nitrate in the water.

Gain of combined nitrogen appears greatly to exceed loss. For instance, in the Mediterranean, which has become depleted of phosphate, rain and rivers have added inorganic nitrogen compounds in excess of the proportion of phosphate absorbed by plants. In consequence this sea is rich in nitrate—Bernard (1939) found some 50 mg. at the surface and 400–600 mg. nitrate-N per m.³ at a depth of 350 m. This compares with some 10 mg. at the surface and 150 mg. nitrate-N per m.³ at 350 m. depth in the eastern Atlantic in the same latitude.

Absorption by plants. All three inorganic nitrogen compounds, ammonium, nitrate and nitrite, can be absorbed by phytoplankton, or at least by some species. There is a very marked preferential absorption of ammonium (fig. 31).

When the organisms become nitrogen-deficient, and are supplied with a nitrogen source, they absorb ammonium and nitrate in the dark, converting them into organic compounds including chlorophyll. Nitrite cannot be utilized in the dark (Ketchum, 1939; Harvey, 1953).

Amino acids of low molecular weight are absorbed as such by at least some species of phytoflagellates (Schreiber, 1927; Braarud and Føyn, 1931), but on present evidence amino nitrogen appears not to be utilized by diatoms without prior deamination. Some amino acids, but not all, set free ammonium rapidly when added to a diatom-bacteria community in culture (Harvey, 1940). In the sea, bacteria are found attached to

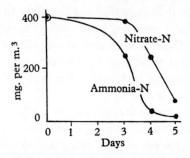

FIG. 31. Fall in concentration of nitrate- and ammonium-nitrogen due to the growth of phytoplankton community.

diatoms and may well act as commensuals permitting organic nitrogen to be utilized.

Urea is considered to dissociate into ammonium and cyanate ions at great dilution (Cooper, 1937), but it does not distil as ammonia under reduced pressure at pH 9 (J. Riley, 1953). With urea as a nitrogen source, growth of diatoms in culture is less rapid than with nitrate.

Uric acid allowed slow growth of a phytoplankton community, perhaps without prior breakdown (Harvey, 1940). Trimethyl-amine oxide, an excretion product of fish, was not utilized as nitrogen source in these experiments with a phytoplankton community.

The proportion of nitrogen to phosphorus in phytoplankton is not a fixed ratio. Cells can be 'deficient' in either and the ratio varies with the relative concentration of each in the medium

(Ketchum, 1939*a*, *b*). Analyses of diatoms collected in the English Channel gave the following values (Cooper, 1937):

	Ratio of N/P by weight
English Channel, mixed diatom community, 3 April 1934	9·6
English Channel, mainly *Rhizosolenia*, 15 May 1934	7·7
English Channel, mainly *Rhizosolenia*, 24 May 1934	7·0

Direct experiment with a mixed culture indicated that some nine times more nitrogen than phosphorus was used (Harvey, 1940).

Discharge from plants. When phytoplankton or animal tissue is voided by animals, the soluble and particulate nitrogen compounds are broken down by bacteria, yielding ammonia. These processes are discussed in the next chapter. However, a residuum of dissolved organic nitrogen remains in the water, perhaps because it is so dilute that the threshold is approached below which bacteria cannot absorb sufficient food to more than counterbalance their rapid respiration.

When plant or animal tissue is digested and excreted by marine animals, more than half their nitrogen is excreted as ammonia.

The following representative figures (Baldwin, 1949) show (i) the percentages of the total nitrogen excreted in various forms by invertebrates (they were obtained by averaging the results of analyses of the excreta of seven different aquatic species), and (ii) the analysis of the urine of a marine teleost, *Lophius*:

	(i)	(ii)
Ammonia	52·2	56
Urea	6·1	5·7
Uric acid	1·3	0·2
Amino acids	14·0	0·0
Creatine	—	6·5
Trimethylamine oxide	—	28·2
Undetermined	26·4	3·4

TABLE 7. *Nitrate + nitrite-nitrogen in mg. N per m³ in the water at positions along the 30° W. meridian of longitude during April–May 1931*

(From data published in *Discovery Reports*, vol. 4.)

Depth (m.)	57° 36′ S.	53° 33′ S.	46° 43′ S.	38° 10′ S.	21° 13′ S.	3° N.	9° N.	14° 27′ N.
0	510	—	210	60	2	2	2	3
20	500	—	210	36	0	1	2	1
40	490	490	210	42	0	1	2	1
60	500	490	230	30	0	2	0	1
80	510	490	240	42	0	29	27	1
100	530	500	250	89	1	63	47	28
150	530	510	310	101	1	170	200	170
200	530	530	380	131	7	220	210	200
400	530	540	450	260	100	230	220	210
800	530	540	480	310	240	220	250	240
1500	510	540	450	360	170	220	210	210
2500	510	530	420	330	180	220	210	210
3200	500	—	—	—	—	—	—	—
3500	—	510	450	320	170	—	210	210
4500	—	—	440	—	—	—	210	210
4900	—	—	—	—	280	—	—	—
5300	—	—	—	—	—	—	—	210

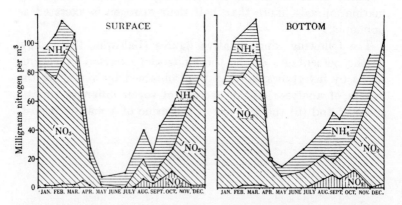

FIG. 32. Variation in ammonium-, nitrate- and nitrite-nitrogen in the upper layers and near the bottom at a position in the English Channel, 20 miles south-west from Plymouth, during 1931. (After Cooper, 1933.)

Distribution of combined inorganic nitrogen.

In the deep water of the Atlantic, where almost all is present as nitrate, its distribution is in general similar to that of phosphate. The greatest concentrations are found around the 800 m. level in high latitudes. The concentration in the Atlantic deep water more than doubles as it moves south to the Antarctic (see table 7). The rate of increase is greater than that of phosphate, perhaps due to slow decomposition of some of the 250 mg. per m³ of dissolved organic nitrogen found by Krogh in the deep water of the North Atlantic as it moves southward.

The seasonal changes in the three inorganic forms in the English Channel are shown in fig. 32.

CHAPTER IV

CHANGES IN COMPOSITION DUE
TO BACTERIA

MANY of the oxidations taking place in sea water are usually attributed to bacteria, who either build up dissolved organic matter into their tissue and excrete it as remineralized products of their rapid metabolism or excrete enzymes which catalyse oxidations in their environment.

In addition, they probably excrete into the water substances which affect the growth of at least some species of plants, such as the vitamin B_{12} (Lochhead, Burton and Thexton, 1952); also they probably adsorb on their surfaces many of the micro-constituents of sea water.

THE BACTERIAL FLORA

The sea contains bacteria freely suspended in the water, attached to organisms, to particles of organic debris, and in the superficial layer of the sea floor where they are abundant. In mud or sand the numbers decrease with depth below the superficial layer, many spore-forming species being present (Waksman, Reuszer, Carey, Hotchkiss and Renn, 1933; Zobell, 1946).

Benecke (1933) and Zobell and Upham (1944) give a very complete bibliography of the species of marine bacteria which have been isolated and described, some from sea water and others only from bottom deposits. Many of these species have been found to bring about chemical changes in the culture media, but it is improbable that all of them act in the same manner in the sea. To do so, some of the species are known to require the conditions of culture, such as plentiful food or ample supply of some constituent which is only at great dilution in the sea.

The population density of bacteria decreases on passing from inshore waters to the open sea. In the open ocean the greatest population is found where phytoplankton is abundant, and in the water immediately above the sea floor, particularly over a sand or shell bottom. Observations by Waksman, Reuszer,

Carey, Hotchkiss and Renn (1933) point to the development of phytoplankton in the sea being closely accompanied by the development of planktonic bacteria. The plants are grazed by animals which do not fully digest the cell substance when food is plentiful, and thus nutrient material is excreted into the water.

TABLE 8. *Some changes brought about through the agency of marine bacteria in the sea or in culture*

Decomposition of chitin
   cellulose
   petroleum hydrocarbons
   lignin
   urea

Oxidation of ammonium to nitrite
   nitrite to nitrate
   sulphide to sulphate

Reduction of nitrate to nitrite
   nitrate to nitrogen gas
   sulphate to sulphide (anaerobic)
   sulphides to sulphur

Utilization of gaseous nitrogen
   ammonium and urea
   nitrate and nitrite
   phosphate

 as source of nitrogen and phosphorus in the bacterial cell, and liberation of these as ammonium and phosphate after death

Formation of concretions consisting of iron and manganese oxides

In the Mediterranean, Bertel (1912) records that bacteria increase in numbers with depth, while at the surface they are mostly killed by strong sunlight to be replaced by others during the night. Vaccaro and Ryther (1954) found that sunlight had no effect upon the respiration of marine bacteria suspended in bottles ten inches below the sea surface.

In clear inshore waters several hundred or more bacteria per * cubic centimetre, comprising twenty-five to thirty-five species, are usually found. In the open ocean away from land the number may drop to less than ten per cubic centimetre where phytoplankton is sparse.

The bacteria which persist in suspension in sea water utilize nutriment at great dilution, there being no more than 2–5 mg. of available organic matter per litre. They are also adapted to growth in the relatively high concentration of salt. Table 9 indicates that only some 10 % of bacteria from the open ocean will live in fresh water, while many terrigenous bacteria which have become adapted to salt water are to be found close inshore. The majority of fresh-water bacteria soon perish in the sea. Great numbers of coliform bacteria are discharged in sewage but are not found in the sea more than a few miles from the outfalls. Pathogenic bacteria have only a short life in sea water; however, if typhoid bacilli are filtered out from the water by shellfish, they may remain alive for some weeks in the fluids within the shell.

TABLE 9. *Relative numbers of bacteria from marine and terrestrial sources which developed in a nutrient medium prepared with sea water and fresh water*

(From Zobell, 1941.)

Source of inocula	Sea water + nutrients	Fresh water + nutrients
Raw sea water	100	9
Inshore water, terrigenous pollution	100	97
Tap water	4	100
Inland soil	15	100
Soil near sea	48	100
Sewage	13	100

Ketchum *et al.* (1949, 1952) have investigated the death-rate of coliform bacteria added to raw sea water. Their death-rate is much reduced if the water is first autoclaved, pasteurized or chlorinated; it is increased if the water is stored previous to the addition of coliform organisms. Addition of organic matter to the raw sea water reduces the death-rate, but to a lesser extent than sterilizing. The production of coliform bactericidal substance by the natural bacterial flora is indicated. Guelin (1954) has found *coli* and *Salmonella* bacteriophages in the harbour water and in the sea off Plymouth.

A variety of methods has been employed to estimate the total number of bacteria in sea water. An extensive study of the number of cells and of species which grow in a variety of nutritive media has been made at the Scripps Institution, with

the object of obtaining maximum plate or dilution counts
(Zobell, 1941, 1946). It was found that most of the bacteria in
sea water and in bottom deposits, at positions remote from the
influence of land drainage, have specific salt requirements which
are best satisfied by natural sea water. Neither synthetic sea
water containing all the major constituents (p. 137) nor other
isotonic salt solutions are satisfactory substitutes. The further
addition of iron to natural sea water led to increased bacterial
development, in some cases to an increase amounting to 76 %,
and to the development of a greater number of species. There
were indications that the addition of phosphate led to slightly
greater development. Successive dilution-method counts using
sea water to which had been added 0·5 % of bacto-peptone and
0·01 % of ferric phosphate gave values 12 % higher than plate
counts in the same medium with 1·5 % bacto-agar. A considerable variety of nutrients was tried and found to give either
inferior or no higher counts than the above. The optimum pH
range lay between 7·5 and 7·9.

During the course of this investigation it was observed that
bacteria behave differently in sea water collected from different
positions, enriched with nutrients and sterilized at 120° C. These
differences in growth-promoting properties disappeared if the
water was first stored at room temperature for a few weeks, presumably because the organic fractions which are responsible for
the differences are oxidized by the increased bacterial activity
accompanying the storage of water in glass bottles.

With regard to other methods of estimating the bacterial
population in samples of sea water, direct observation is difficult,
since it is necessary to concentrate the cells. It yields higher
counts than are obtained by plate or dilution methods (Cholodny,
1929), and presumably includes some cells which do not grow
in the nutritive media employed in the culture methods and
some dead cells which have not completed autolysis.

When a glass slide is hung in the sea, bacteria rapidly attach
themselves to the surface; the number of cells which attach
under standardized conditions bears a direct relation to plate
counts (Hotchkiss and Waksman, 1936).

The temperature for most rapid multiplication lies in the
neighbourhood of 20° C. for most marine bacteria (Zobell and

Upham, 1944). These authors also state that ability to grow at subzero temperatures is a common property of most marine bacteria, and that their thermal death-point is considerably below that of most terrigenous or fresh-water bacteria. Ten minutes at 30° C. killed about one-quarter of the bacteria collected from cold deep ocean water, and 10 minutes at 40° C. killed more than three-quarters.

EFFECT OF STORAGE ON BACTERIAL FLORA

The study of changes taking place in sea water under natural conditions in the oceans is hampered by the remarkable events which happen when sea water is stored in a glass vessel. Whipple reported in 1901 that when tap water was put into glass bottles the number of bacteria fell during the first 3–6 hours by 10–25 % and later increased by many hundred per cent, with a reduction in number of species. This rise in bacterial numbers was several times greater in small than in large bottles, and was reduced or even nullified if the water was kept agitated. A similar rise in bacterial numbers takes place when sea water is stored in glass vessels. Waksman and Carey (1935) found that multiplication took place in Seitz and in colloid-filtered water which had been inoculated with raw water, and from the oxygen used calculated that the rapid growth of bacteria breaks down about one-third of the organic matter in solution. Zobell and Anderson (1936 a) and Lloyd (1937) found a much greater increase in numbers of bacteria when the water was stored in small than in large bottles, or in bottles where the water-glass surface had been increased by filling the bottle with glass beads or rods. By observing the numbers at close intervals of time, Miss Lloyd found that the increase followed the course of a population curve; the peaks were determined by the volume of the container, but were not affected by the surface area of the water exposed to the air. Zobell and Anderson found that the peaks, or maximum number of bacteria found in the water, showed a rough direct proportion to the volume surface area of the bottles, the maxima in small vessels being about twice the maxima in vessels ten times as large. They also observed that, when the surface area was increased by a shallow layer of glass beads or silica grains, the resulting increase in numbers of bacteria was less than that calculated from the volume/area ratio. They concluded that not only this ratio, but also the proximity of the main body of water to the glass surface, played a part. The greatest increases were found in water between sand grains, where populations of some twelve million bacteria per cubic centimetre developed in water which maintained no more than a few hundred bacteria per cubic centimetre in the sea.

Whipple had found that the marked difference between maximum bacterial population, which arose when tap water was stored in small and in large bottles, was much reduced if a small quantity of peptone (5 mg. per litre) was added to the water. Zobell and Anderson noted that if nitrite and 10 mg. per litre of peptone were added to sea water, there was a greater and more rapid loss of nitrite in smaller than in larger vessels, but no such difference when 100 mg. per litre of peptone was added. It appears that the volume effect only occurs when bacteria develop in water containing food substances at very great dilution. This conclusion was also reached by Heukelekian and Heller (1940), who found no growth of the bacterium *Escherichia coli* in solutions containing 0·5 and 2·5 mg. per litre of glucose and peptone, but growth did occur if glass beads were added. With 25 mg. per litre, growth took place without the addition of glass beads, and with concentrations greater than this the effect of adding glass beads faded out.

As Whipple had found for tap water, so Zobell and Anderson found a reduction in number of species when sea water was stored. Some twenty-five to thirty-five species were generally found immediately after the water had been collected, falling to nine or ten species by the time bacterial numbers had reached a maximum and to no more than four or five species after the maximum population had declined. After this decline the population remained relatively high for a long period, the numbers fluctuating from a few thousand to over a hundred thousand per cubic centimetre—a sample of sea water which had been stored at 2–6° C. for 4 years was found to contain 209,000 bacteria per cm^3

Zobell and Anderson (1936) and Waksman & Renn (1936) observed that in full and stoppered bottles the consumption of

5

oxygen continued undiminished for some time after the bacteria in the water had reached maximum numbers and while the population was falling. The former investigators have shown that great numbers of bacteria develop on the glass surfaces; one experiment indicated that within 24 hours more than twice as many were attached to the surface of the glass as were in the water. This accounts for the continued consumption of oxygen after the number of bacteria in suspension have declined.

TABLE 10. *The oxygen content of sea water stored at* 16° C. *in glass-stoppered bottles of different capacities after 20 days, and the maximal bacterial population reached (after 3–5 days) in similar bottles*

The water initially contained 5·46 cm³ O₂ per litre and 231 bacteria per cm³
(From Zobell (1936), *Proc. Soc. Exp. Biol., N.Y.*, 35, 271.)

Volume of sea water (cm³)	10	100	1000	10,000
O₂ per litre	2·59	2·90	3·68	4·17
Bacteria per cm³	1,475,000	1,080,000	673,000	382,000

In Zobell and Anderson's investigation a series of experiments was made dealing with the effect of oxygen on the proliferation of bacteria in stored sea waters. No material effect was observed unless the water was less than 50 % saturated with air. Waksman and Carey, on the other hand, observed greater growth in fully aerated than in partially aerated water. This was confirmed later by Waksman and Renn (1936). Subsequent observations by Zobell (1940) led to the conclusion that the rate of oxygen consumption was independent of the partial pressure of oxygen between 0·31 and 12·74 cm³ per litre at 22° C.

Most of the experimental observations have been made with sea water from which only the larger plankton organisms have been removed, leaving nannoplankton. However, Keys, Christensen and Krogh (1935) and Waksman and Carey (1935) have filtered sea water through collodion ultrafilters, removing all plankton and colloids, and found that about as many bacteria developed in this, after reinoculation, as in the unfiltered water.

Removal of the plankton and bacteria by Seitz or membrane filtration reduces the oxygen consumed after reinoculation, but not always to a very marked degree. It leads to a higher rate of bacterial multiplication, due perhaps to the removal of phagocytic Protozoa.

The proliferation of bacteria when water is enclosed in glass vessels and the effect of their size is attributed by Zobell and Anderson to the water-glass surfaces:

(i) Providing a resting place for attached bacteria, many marine species having periphytic tendencies and at least some being obligate periphytes. In this connexion it is pertinent that saprophytic bacteria are attached to suspended particles in the sea (Lloyd, 1930) and that bacteria are most numerous where plankton is most abundant, as observed by Waksman, Reuszer, Carey, Hotchkiss and Renn (1933), who consider that 'bacteria exist only to a very limited extent in the free water of the sea, but are largely attached to the plankton organisms'.

(ii) Concentrating organic substances from very dilute solution on the surfaces owing to adsorption or other physical attraction.

(iii) Retarding the diffusion of bacterial enzymes away from the cell, where the cell is attached to a solid surface; it has been generally observed that attachment to particles exerts a favourable influence on their enzymatic activity.

Regarding the second suggestion—that organic matter in solution is adsorbed on solid surfaces—the authors state that the accumulation of a film of organic matter on glass slides soon after being submerged in sea water can be demonstrated by differential stains as well as by microchemical technique. Stark, Sadler and McCoy (1938) also state that an accumulation of organic matter can be detected on glass slides which have been immersed for several hours in lake water. Their method of detection was based on the oxidation of a sulphuric acid-dichromate mixture. In neither of these communications is the exact technique described. The writer has been unable to obtain definite results on these lines, but a number of observations have been made which point, indirectly, to adsorption taking place (Harvey, 1941).

With regard to the third suggestion—that bacterial enzymes diffuse from the cell into the surrounding water—the following observations are of interest. Kreps (1934; also Barkova, Bovsook, Verjbinskaya, Kreps and Lukyanowa, 1936) found that changes in ammonium and in nitrate took place in water which had passed a Seitz filter or been sterilized with mercuric chloride. He suggests that sea water, particularly water near the bottom where organic matter is decomposing, contains enzymes which bring about these changes. Keys, Christensen and Krogh (1935) have also observed changes in the ammonium content of water which had been sterilized with mercuric chloride, while Newcombe and Brust (1940) have noted that saturating water with chloroform reduces but does not stop phosphate being set free during storage. Cooper (1937) has also found changes in the ammonium content of water sterilized with mercuric chloride.

Wood (1951, 1953) discusses the difficulty of explaining why the small number of bacteria which can be detected in the sea bring about the transformations usually ascribed to them. He suggests that the bacteria do not act directly as much as by 'producing catalysts or by creating conditions for reactions which catalysts already present may carry out'.

It is an outstanding question why offshore sea water, which contains sufficient nutriment for the production of several million bacteria per cubic centimetre, and will in fact rapidly produce this population when in contact with clean sand grains, normally supports a population of no more than 10–200 bacteria per cm^3 as free-living planktonic cells. In addition to lack of solid surfaces, Protozoa and other animals keep the bacterial fauna eaten down; this has been stressed by Zobell and by Waksman and Hotchkiss (1937). The former investigator (1936) has also concluded that natural sea water contains a bacteriophage, or heat-labile substances inimical to the growth of bacteria; added bacteria grew more rapidly in autoclaved than in Berkefeld filtered sea water. There is some natural brake upon the growth of planktonic bacteria in the open sea.

CHEMICAL CHANGES TAKING PLACE
DURING STORAGE

Before discussing the changes which take place in sea water during storage it is helpful to consider in broad outline the food requirements and fate of planktonic bacterial cells.

Given suitable conditions and a sufficiency of food, the bacteria continue to build up cell substance and to divide. Their cell substance is singularly rich in both nitrogen and phosphorus. Waksman and Carey (1935) give the ratio of carbon to nitrogen as 5 to 1, while Waksman, Hotchkiss, Carey and Hardman (1938) find that four and a half times more nitrogen than phosphorus is utilized by marine bacteria during growth. These observations indicate a ratio of 45 carbon:9 nitrogen:2 phosphorus in the bacterial cell.

Meanwhile the growing bacterial cell respires, consuming oxygen and liberating carbon dioxide. The rate of respiration varies with different species of marine bacteria (Johnson, 1936), with the concentration of available food (Zobell, 1940) and with temperature. The rate is singularly rapid; thus Zobell calculates that 1 g. of actively growing marine bacteria consumes some 30 cm.^{3} of oxygen per hour at 22° C., whereas marine animals are stated to consume 0·002–1·0 cm.^{3} of oxygen per hour per gram of living tissue.

Provided the environment remains suitable and there is sufficient food, the bacteria continue to respire, build up cell substance and divide.

If the organic matter in solution used as food contains nitrogen or phosphorus in excess of the bacteria's requirements, a part, if not all, of this excess is split off as ammonium or phosphate.

If the organic matter used as food is insufficiently rich in nitrogen or phosphorus, some species obtain their nitrogen requirement from ammonium in solution, some from nitrate or nitrite, while some species can obtain their phosphorus from phosphate in solution.

When the food supply runs out or conditions become unsuitable the cells die and autolyse, giving off into the surrounding water part of their contained nitrogen as ammonia, all or nearly all their phosphorus as phosphate, and much of their contained carbon as carbon dioxide. A small residue may remain which eventually falls to the bottom as 'marine humus'. Once death has taken place autolysis proceeds rapidly. The experiment shown in fig. 20 indicates that all, or almost all, the phosphorus in the bacterial cells was returned to the water within a week after the food supply for the bacteria had run short. Besides death and autolysis, many bacterial cells doubtless end their existence by being eaten and digested by protozoa and filter-feeding animals.

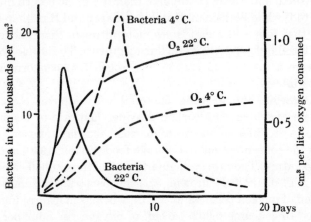

FIG. 33. Change in bacterial population and consumption of oxygen in stored sea water at 4 and 22° C. (After Waksman and Renn, 1936.)

A concept such as that outlined in the preceding paragraphs throws light upon the changes which take place during the storage of sea water, or of sea water to which dead plankton or organic compounds have been added.

The change in oxygen content of raw sea water, containing small plant and animal organisms as well as bacteria, has been followed by several observers for storage periods up to about three weeks. A rapid consumption of oxygen and proliferation of planktonic bacteria is followed by a slower consumption, due mostly to bacteria adherent to the glass (fig. 33). The oxygen consumption during the first five days, when fresh or sea water is stored, is usually about half the consumption at the end of three weeks. After this, consumption was found by Waksman and

Renn (1936) to continue slowly in two experiments lasting nine weeks. Rakestraw (1947) found little consumption after 50 days in water from mid-depths and after 100 days in surface water, in experiments where waters were stored for long periods.

The quantity of oxygen consumed by raw water, collected from different positions, when stored in the dark at the same temperature, varies considerably. The samples of water will contain varying quantities of minute plants and animals; there is evidence that the dissolved organic matter also varies. Relatively low values of oxygen consumption in offshore waters are recorded—values of $c.$ 0·5 cm.^{3} O_2 per litre, compared with over 1 cm.^{3} consumed by inshore water which had been filtered free from all organisms and reinoculated in experiments by Waksman and Carey (1935).

In general, samples of inshore water consume more oxygen than samples from the open sea, where water from greater depths consumes more oxygen than water from the upper layers (Seiwell, 1937). Temperature plays a predominating part in controlling the rate of consumption. Johnson (1936) records Q_{10} for washed marine bacteria varying between 2·2 and 2·3, values which agree with observations by Seiwell (1937), who gives the following data.

Average oxygen consumption of all samples stored in the dark:

at 24° C.,   1·224 cm.^{3} O_2 per litre,
at 11° C.,   0·487 cm.^{3} O_2 per litre.

The rate of consumption in raw water from the upper layers of the sea is sometimes reduced through insufficient phosphorus for the bacterial requirements, and may be increased by the addition of phosphate (Keys, Christensen and Krogh, 1935). There is sufficient available nitrogen for the bacteria; addition of ammonium has no effect.

The change in pH due to carbon dioxide set free in raw waters has been followed by Atkins (1922). The least change corresponded to the production of 0·9 mg. C per litre—about one-half of the total organic carbon thought to be present as organic compounds in solution in ocean water.

An increase in phosphate during storage of raw water has been recorded by several observers, sometimes preceded by a decrease due to utilization by bacteria where the organic food contains insufficient phosphorus.

The ammonia in solution usually decreases when sea water is stored in the dark; in some waters an increase has been observed (Keys, Christensen and Krogh, 1935; Cooper, 1937a). Such changes also take place in waters sterilized with mercuric chloride or by filtration. Cooper has observed a reduction in ammonia after 13 days' storage in a series of samples from the open sea, with no corresponding increase in nitrite. He considers that ammonium is oxidized to hyponitrite, the reaction being activated by some agency which is not necessarily bacterial since sterilization does not stop it.

Increases in the nitrite and nitrate content have not been found in water collected from offshore, although looked for by several observers, unless contaminated with bottom deposit (see Carey, 1938). This is odd, because the distribution of nitrite and ammonia in the oceans indicates that both nitrite and nitrate are formed in the layer of water poor in oxygen below the photosynthetic zone. Two possibilities suggest themselves: either the nitrite- and nitrate-forming bacteria die out during storage of more or less plankton-free water (Harvey, 1941), or these bacteria are only active when attached to plankton. If dead plankton is added to raw water, both nitrite and later nitrate are slowly formed (Carey, 1938; Von Brand and Rakestraw, 1937–42).

BACTERIAL DECOMPOSITION OF ORGANIC COMPOUNDS AND OF DEAD PLANKTON ORGANISMS ADDED TO SEA WATER

When a bacterial food devoid of nitrogen, such as a sugar, is added to sea water, the rate and quantity of oxygen consumed, due to the growth of bacteria, is increased. The available nitrogen in the water, added to that liberated by decomposing nitrogenous organic matter in solution, allows part of the added sugar to be utilized rapidly.

Waksman and Carey (1935), in experiments with an inshore water which consumed 1·3 cm.^{3} of oxygen per litre in five days, found that the oxygen consumption was doubled on adding

2·5 mg. per litre of glucose, but that further additions of glucose caused no further additional consumption unless a source of nitrogen was also added (fig. 34). They made two inferences from this and other experiments: (i) since the oxygen consumption due to 2·5 mg. of added glucose was equal to the oxygen consumption of the raw water, the latter contained a similar quantity (2·5 mg. per litre) of readily available organic matter; (ii) since the increased oxygen consumption was two-thirds of the quantity required for the complete oxidation of the

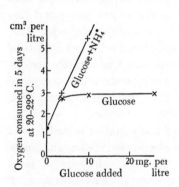

FIG. 34. Effect of adding glucose and of adding glucose + ammonium on the oxygen consumed by bacteria in a sample of inshore sea water.

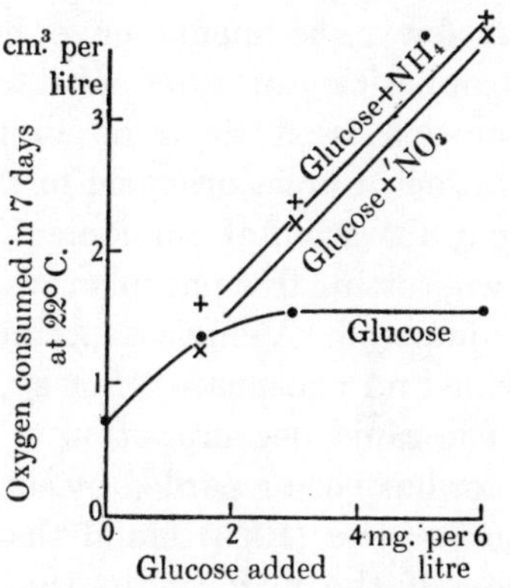

FIG. 35. Effect of adding glucose, glucose + ammonium and glucose + nitrate on the oxygen consumed by bacteria in a sample of inshore sea water.

2·5 mg. per litre of sugar, about two-thirds had been used in respiration and one-third built up into bacterial tissue which remained unautolysed at the end of the five-day period.

A similar experiment with an inshore water (Fig. 35) showed that the addition of 1·5 mg. per litre of glucose doubled the oxygen consumption. Larger additions of glucose led to no greater oxygen consumption in the seven-day period, unless an additional source of nitrogen was provided. Ammonium appeared to be rather more readily utilized than nitrate. The increased oxygen consumption due to the addition of 1·5 mg. glucose was again two-thirds of the quantity required for complete oxidation of the sugar (Waksman and Renn, 1936).

The increased development of bacteria and consumption of oxygen due to the addition of various amino acids has been investigated by Waksman and Renn (1936). In five days at 20° C. between 50 and 70 % of the oxygen required for complete oxidation was consumed. A more detailed study of the breakdown of asparagine has been made by Waksman, Hotchkiss, Carey and Hardman (1938). In from six to nine days at room temperature some 70 % of the oxygen required for complete oxidation was consumed and a similar proportion of the contained nitrogen was liberated as ammonium.

A few experiments have been made concerning bacterial decomposition of other substances added to sea water and the liberation from them of ammonia or phosphoric acid. The addition of urea, uric acid or trimethylamine did not lead to a rapid liberation of ammonia. Nucleic acid was rapidly broken down, setting free phosphoric acid, while casein was less rapidly decomposed. A single experiment with glycerophosphoric acid yielded no phosphate (Harvey, 1940).

The rapid decomposition of dead zooplankton added to sea water has been recorded by several observers. Waksman, Carey and Reuszer (1933) found that at 16–20° C. in 19 days about one-half the nitrogen in the dead animals was liberated as ammonium and about one-fifth of the carbon as CO_2. A similar experiment by Waksman, Hotchkiss, Carey and Hardman (1938) showed rapid liberation of ammonium and phosphate. Seiwell and Seiwell (1938) record rapid initial liberation of phosphate which soon became slower, with marked diminution in size of the particles of detritus. Cooper (1935) has followed the liberation of phosphate from zooplankton in water at 14–19° C. for a period of several months. A rapid initial liberation was followed by a lull; later a second liberation occurred, leading to a maximum phosphate content of the water at the end of two months. The quantity of phosphorus liberated as phosphate was considerably in excess of the quantity in the added zooplankton. This suggests that the plentiful bacterial food allowed the growth of species of bacteria which decomposed phosphorus containing organic substances in solution in the water. When the waters were stored without added zooplankton there was no increase in phosphate comparable in quantity to this excess.

Several studies have been made concerning the decomposition of phytoplankton. Von Brand and Rakestraw (1937–42) added living diatoms, grown in culture, to sea water, and stored this in the dark for several months, keeping it aerated. Initial analyses gave the quantity of nitrogen in the added diatoms and any particles of organic matter present in the water. After about one month—the period varying with the temperature for the most part—a quantity of ammonium-nitrogen had been set free into the water, equal to or exceeding the nitrogen in the particulate matter at the beginning of the experiment. Where, as usually happened, the quantity exceeded that in the added

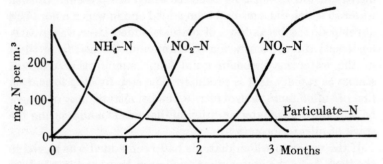

FIG. 36. Production of ammonium, followed by its conversion to nitrite and finally to nitrate, in sea water containing diatoms while stored in the dark. (After Von Brand and Rakestraw.)

diatoms and suspended particles, it may be inferred that the excess was derived from organic nitrogen compounds in solution in the water. The ammonium set free changed to nitrite and subsequently to nitrate (fig. 36), except under anaerobic conditions when the sequence stopped at the ammonium stage.

The rate of ammonium formation was about doubled for a rise of 6–8° C., and in most experiments the subsequent oxidation to nitrite and nitrate was affected by temperature in a similar manner except at low temperatures in the region of 1–2° C., when little or no change took place. In one experiment where water collected from a depth of 1200 m. was used, this subsequent oxidation was exceptionally slow, presumably due to the slow growth of nitrite- and nitrate-forming bacteria. It is of interest that Zobell has observed that bacteria grow at different rates in waters collected from different positions and similarly enriched with nutrients, suggesting that sea water contains some growth-promoting factor or *trace* element affecting bacteria and that such occurs in variable quantity. A similar inference arises from experiments on the growth of diatoms (p. 107).

A series of experiments, in which diatoms were added at intervals to sea water stored in the dark, showed that after the second addition the production of ammonium, nitrite and nitrate may take place simultaneously, the predominating product of bacterial action depending upon the dominance of ammonium-, nitrite- or nitrate-forming bacteria which are present. In such water to which diatoms had been added and in which a flora had developed, the breakdown of diatoms added later was greatly hastened; the concentration of ammonium formed from them in the water may remain quite low, being further oxidized almost as rapidly as it is produced. The step-by-step formation first of ammonium, then of nitrite, then of nitrate, may no longer be apparent if the later addition of diatoms is made during the stage of nitrate formation.

If the water to which diatoms had been added was stored in the dark until the ammonium maximum had been reached, or until any subsequent stage, then reinoculated with a few living diatoms and exposed to the light, all the ammonium nitrite or nitrate was used by the growing diatoms. Subsequent storage in the dark led to the regeneration of ammonium, followed by its conversion, step by step (fig. 37). In this manner regeneration of organic nitrogen to nitrate, followed by its utilization, was repeated in the same culture three times. It is odd that the regeneration occurred step by step, there being a marked lag before nitrite and later nitrate formation took place, in spite of the rich flora of nitrate-forming bacteria present at the times of reinoculation.

In connexion with these investigations Von Brand (1938) determined the organic nitrogen in particulate matter filtered from samples of sea water. He found only 5–18 mg. N per m³ in the upper layers of the sea except where phytoplankton was very abundant, and in the deeper water below 300 m. no more than 1–3 mg. per m³ at five positions in the open ocean.

The liberation of phosphorus in an available form was not followed in these experiments, but it presumably took place at the same time as ammonium formation, since there was enough there for the diatoms added as a reinoculation to grow (fig. 37). However, there is no data of the quantity of phosphate in the water at the start of the experiments.

Dried diatoms when added to sea water were found by Waksman, Stokes and Butler (1937) to liberate two-thirds of their phosphorus as phosphate in three weeks at 22° C. Living

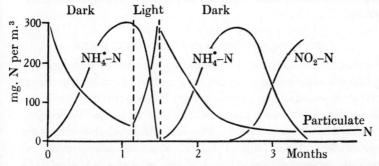

FIG. 37. Production of ammonium in sea water containing diatoms while stored in the dark, the utilization of this ammonium while kept in the light, and the subsequent production of ammonium and its conversion to nitrite in the dark. (After Von Brand and Rakestraw.)

diatoms were observed to be relatively resistant to bacterial decomposition until they have been attacked by Protozoa. Cooper (1935) stored in the dark sea water to which living diatoms had been added. After an initial lag period phosphate was liberated, but even after three months no more than two-thirds of the added organic phosphorus had appeared as phosphate.

BACTERIAL OXIDATION OF AMMONIUM TO NITRITE AND NITRATE

The various observations and experiments already mentioned indicate that bacteria which oxidize ammonium occur on plankton organisms, in bottom deposits and in the water immediately above the bottom, but that they do not occur freely suspended in the main body of the water away from land and away from the immediate influence of the bottom.   ★

There can be no doubt that the water close to the bottom is a site of active nitrification, particularly in waters of moderate depth. Experiments by Zobell (1935b) indicate that nitrifying bacteria which had been from isolated bottom deposits were ideally suited to oxidize ammonia in the sea water but not within the deposits. It was found that the optimum oxidation-reduction potential for nitrification lay within 0·35–0·55 V.; the potential of sea water is c. 0·45 V., due to the irreversible oxygen system (Cooper, 1937b), far removed from the low potentials found to exist in bottom deposits (Zobell and Anderson, 1936b).

FIXATION OF ATMOSPHERIC NITROGEN

Bacteria of the aerobic *Azotobacter* species have been isolated by several observers and found attached to plant organisms, in bottom deposits and free in the water. The anaerobic *Clostridium* species has also been found in bottom deposits. Waksman, Hotchkiss and Carey (1933), who have isolated these bacteria from waters of the Gulf of Maine, conclude, after reviewing previous investigations, that the sea contains an abundant population capable of fixing appreciable quantities of nitrogen. They require a relatively rich supply of food. It still remains to be determined to what extent this process takes place in the sea, if at all.

NITROGEN-LIBERATING BACTERIA

Bacteria which are capable, under suitable culture conditions, of reducing nitrates and nitrites to nitrogen have been found in bottom deposits and in the sea by several observers. In order to set free gaseous nitrogen, they require readily available food as a source of energy and the presence of a relatively high concentration of nitrate; it is doubtful if these conditions are fulfilled in the oceans. A hypothesis, advanced by Brandt in 1899, that the quantity of nitrogen-containing salts in the sea was controlled by the activities of such bacteria, has received much attention. Subsequent considerations advanced by Gran, by Issatchenko and by Waksman lend no support to this hypothesis. Waksman, Hotchkiss and Carey (1933) have reviewed the various investigations on this subject. Thompson and

Gilson (1937) quote experiments by Cranston and Lloyd on a bacterium isolated from the sea. These yielded results which can be explained by the assumption that nitrate was completely reduced to nitrite before the latter was reduced to hyponitrite, each stage being completed before the next stage commenced and before nitrogen was finally set free in the culture.

NITRATE-REDUCING BACTERIA

In addition to true denitrifying bacteria which can reduce nitrate to atmospheric nitrogen, bacteria which are able to reduce nitrate to nitrite are abundant in the sea and in bottom deposits. In order to carry out this reduction, they also require a relatively rich supply of food. Waksman, Reuszer, Carey, Hotchkiss and Renn (1933) incubated a series of water samples from various positions and depths in the Atlantic after adding nitrate, phosphate and calcium acetate as source of food. Many but not all were found to produce nitrite.

Thompson and Gilson (1937) considered that the nitrite-rich layer occurring below the photosynthetic layer in the Indian Ocean was probably due to the reduction of nitrate through the agency of such bacteria. The later observations by Von Brand and Rakestraw, that dead diatoms yielded ammonia which was subsequently converted to nitrite and then nitrate, suggest that this nitrite-rich layer may be due to nitrification by bacteria attached to plankton organisms, rather than to denitrification. It is perhaps pertinent that several observers have noted that nitrite is produced in cultures of diatoms growing in water rich in nitrate; Zobell (1935) considers the possibility of extracellular reduction of nitrate by the plants. There is no clear evidence showing whether the nitrite in the nitrite-rich layer of the oceans is produced by nitrification of ammonium or denitrification of nitrate.

The reduction of nitrate to ammonium in the sea can be brought about by bacteria which utilize nitrate when their food supply is short of organic nitrogen and subsequently give off ammonium during autolysis. Whether there are species which bring about reduction by another mechanism, denitrification, is open to some doubt (Waksman and Carey, 1935).

CHAPTER V

FACTORS INFLUENCING THE GROWTH OF PLANTS

INTRODUCTION

IT is the aim to gather in this chapter evidence which bears upon the effects of several factors on the plant organisms. Before considering how each factor may affect a single plant, it is helpful to consider how the tangle of variables affects the whole population of marine plants in nature, a population which waxes and wanes, changing its constitution through the seasons and differs both in constitution and in quantity from one sea area to another.

The *standing crop* is a momentary balance resulting from the rate at which the plants have been growing and the rate at which they have been eaten. The *growing crop* is that portion which is present within the photosynthetic zone, the depth of which changes during the day and from day to day.

The rate of production of organic matter below unit area of the sea, at any moment, is the rate of photosynthesis by the growing crop minus its rate of respiration. Most of the marine plants increase in numbers by simple division, the daughter cells then build up their organic contents and divide again. Two processes take place, production of organic matter and division. Under certain circumstances the capacity to divide can be impaired, resulting in loss of buoyancy and finally death.

The increases and decreases of the standing crop in the sea appear, on present evidence, to be mostly due to changes in the following environmental conditions:

(i) Incident light, which in conjunction with turbidity of the water determines the depth of the photosynthetic zone and also the distribution of light intensity within the zone.

(ii) The rate of replenishment of available nitrogen and phosphorus in the photosynthetic zone. This in conjunction with the rate of utilization determines their concentrations, which reduce the rate of plant growth if either fall below a threshold value.

This rate of replenishment depends upon turbulence (vertical eddy diffusion) and in some areas also upon upwelling of water to take the place of surface water moving away as a current. Replenishment is also taking place from the phosphate and ammonia excreted by zooplankton while they are present within the photosynthetic zone.

Replenishment of phosphate is continuously taking place by hydrolysis of dissolved organic phosphorus compounds. It is likely that replenishment of ammonia is also taking place due to deamination of dissolved organic nitrogen compounds.

(iii) The rate at which the zooplankton are eating the plants.

(iv) The rate at which the plants are being carried down by turbulence below the photosynthetic zone, thereby causing diminution of the growing crop. In this respect turbulence acts in opposition to its beneficial effect of replenishing nutrients.

In addition to loss of the growing crop by being eaten and by turbulence, under some undetermined conditions a dominant species of the phytoplankton may lose buoyancy and sink.

(v) The effect of increased temperature is to increase the growth rate of plants while illuminated, and to increase loss by respiration while in the dark. At low light intensities such as occur in the lower part of the photosynthetic zone, there is some evidence that temperature may have little or no effect upon growth rate.

In addition to affecting plant growth, seasonal changes in temperature play a part in affecting the population of zooplankton grazing upon the plants.

With the exception of temperature, which is only subject to steady slow change, each one of these sets of factors has been observed on occasions to play a part, and even an overriding part, in affecting the plant population.

The foregoing remarks apply to the population density of phytoplankton organisms as a whole. The causes of differences in species composition, which are found from sea area to sea area and through the seasons, are obscure. Temperature doubtless plays some part.

When dissolved silicate is at, or becomes reduced to, low concentrations, diatoms give way to species without siliceous skeletons.

6

When available nitrogen and phosphorus is at, or becomes reduced to, low concentrations, the larger species tend to be replaced by smaller plants (see p. 97).

There are some species which are only found close to the coast in any abundance, but many appear in both oceanic and inshore positions. Corlett (1953) has followed the seasonal changes in diatom species at two oceanic positions, 54° N., 20° W. and 60° N., 20° W., finding that the majority of species which attain abundance are those which are often abundant in the inshore waters of the English Channel.

ILLUMINATION

The quantity of light penetrating the surface and illuminating the upper layers of the sea each day, and the change in intensity with increasing depth, play a leading role in determining the growth of phytoplankton.

The light falling on land or sea is ordinarily measured in metre-candles (1 metre-candle = 1 lux). This is an arbitrary unit of intensity of white light giving the same response to the eye as a standard reproducible source of white light at a specified distance. If the standard source emits light with a different spectral composition from the light being investigated, intensities of the same candle power will not have the same flux of energy. If the light being investigated changes in spectral composition, the flux of light energy per unit candle power changes also.

At noon the direct sunlight (of wavelength 360–760 mμ, which contains about half of the total energy of the radiation from the sun) has a different composition to the light from the sky, and both change in composition as the sun approaches the horizon. Thus the light falling on the sea is continually changing in spectral composition.

Sunlight at noon with an intensity of 1 kilolux has an average flux of energy of 0·0006 g.cal. per cm.^{2} per min. The lux/energy relation of the light falling on the sea during the day varies somewhat from this value.

On a sunny summer's day in latitude 50° N., with some white clouds in the sky reflecting light from the sun, the intensity of light from sun and sky may reach 130 kilolux, and, if the sun is

obscured, about half this value. On a sunny winter's day it may
reach some 25 kilolux at noon, and on a cloudy winter's day
some 5 or 6 kilolux.

TABLE 11. *Relative energy distribution in the
spectrum (in percentages)*

6.00 millimeter

Light	Wavelength (mμ)	Mean noon sunlight* (%)	Blue sky light† (%)
Violet-blue	380–490	26·7	50
Green	490–560	24·2	24·3
Yellow and orange	560–620	19·7	12·3
Red	620–720	29·4	13·4
		100·0	100·0

* From data by Abbott.
† From data by Walsh and by Poole and Atkins; quoted by Jenkin (1937).

Throughout the year in the tropics, except during a rainy
season, and in latitude 50° N. during May and June, the average
quantity of light amounts daily to some 600 kilolux hours, an
average intensity over the whole 24 hours of 25 kilolux, with
an average flux of some 0·15 g.cal. per cm² per min. During
December and January the quantity falls to about one-ninth of
600 kilolux hours in latitude 50° N. The monthly averages vary
from year to year (see fig. 38).

On reaching the sea a variable quantity is reflected, about
4 % in sunny calm weather—more when the sun approaches the
horizon—and as much as 25 % with moderate winds and the
sun shining. On an overcast day about 8 % is reflected.

As light rays penetrate the sea, they are both absorbed by the
water and by particles suspended in it, and, in addition, are
scattered. Rays are deflected by molecules of water (molecular
scattering) and reflected in all directions by organisms and
detritus in suspension.

Blue light is absorbed less and scattered more than red light.

Looking down into the sea far from land it appears blue,
owing to blue light being scattered upwards. As land is approached it appears green, there being more particles in sus-

6-2

pension which reflect light of all wavelengths upwards. The water always contains organisms and a material quantity of inorganic particles in suspension. (Observations some 100 miles from land in the north-east Atlantic show about 0·1 g. per m³, with rather more at the surface, while 20 miles offshore in the English Channel around 1 g. per m³ is found. Variation with depth and position in the oceans of light scattering has been investigated by Jerlov, 1953.)

Furthermore, in coastal water, notably in the Baltic, there is a trace of dissolved carotenoids—'Kalle's yellow substance'—which absorbs blue and near ultra-violet light.

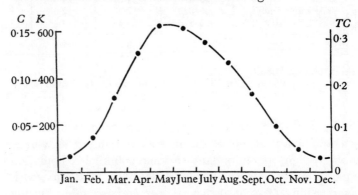

FIG. 38. Average illumination at Plymouth for the years 1930–7. K = kilolux hours per day; C = g.cal. per cm² per min., for light within the range 380–720 mμ wavelength. TC = deduced total solar radiation, which includes infra-red, in g.cal. per cm² per min. (Data from Atkins, 1938.)

Due to the combined effect of absorption and scattering, the spectral composition of the light changes with increasing depth; if the suspended particles do not change in quantity with depth, the intensity of flux of light of any one wavelength decreases exponentially. Then

$$\text{the extinction coefficient } k = \frac{2 \cdot 3(\log_{10} I d_1 - \log_{10} I d_2)}{d_2 - d_1},$$

(the extinction coefficient k includes both light absorbed and lost by scattering), where $I d_1$ is the intensity at depth d_1 metres and $I d_2$ at d_2 metres.

In the sea the rate of absorption of light is usually somewhat greater in the surface layers than in the deeper water below; at great depths in the oceans a layer of more transparent water has been observed close to the sea bed.

Fig. 39 shows how the mean coefficient and percentage transmission per metre vary with wavelength.

In the extreme case of turbid harbour water, the effect of scattering exceeds that of absorption; less blue-green and blue light is transmitted than red light, which penetrates the farthest.

The rate at which different wavebands of light decrease with depth—their extinction coefficients—have been observed by

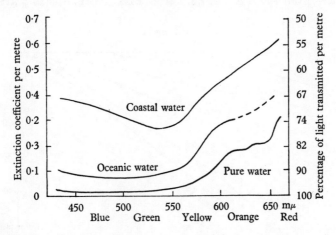

Fig. 39. The extinction coefficient per metre and the percentage transmission of light of different wavelengths in sea waters. (After Sverdrup, Johnson and Fleming, 1942.)

lowering a photometer covered with a light filter which lets one particular waveband pass.

Fig. 40 shows the diminution with increasing depth of such wavebands as observed in the English Channel. Knowing the rates of decrease for the several bands and knowing the energy flux of the light entering the sea and the proportion of energy in the different bands, it is possible to calculate the total energy of light arriving at different depths. This has been done by Jenkin (see p. 87) in studying the rate of photosynthesis by plants at different levels.

A rough method of comparing absorption of light in different sea areas is to lower a white disk, 30 cm. in diameter (a Secchi disk) until it is no longer visible. In moderately clear water it has been found that some 16%, more or less, of the incident light penetrates to the depth in metres at which the disk is no longer visible. This depth roughly approximates to

$$D = \frac{1 \cdot 7}{\text{extinction coefficient}}$$

There are subjective errors of vision; also, whereas forward scattered light arrives on the disk from all directions, there is a loss from the beam of light from disk to eye, due to scatter.

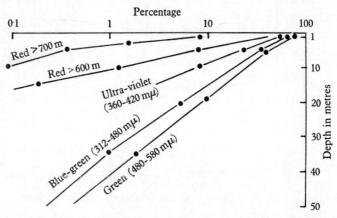

FIG. 40. Diminution of intensity with depth of light falling on the surface of the sea. English Channel, September 1936. (After Poole and Atkins, 1937.)

Light intensity and the growth of phytoplankton. The effect of light intensity upon oxygen production by diatom cultures, exposed in bottles suspended in the sea at different depths, has been investigated by several workers. The increase in oxygen in the bottles is that produced by photosynthesis, less the quantity used in respiration by the plants and bacteria. If the bottles are filled with raw sea water instead of a culture, respiration by planktonic animals, as well as by bacteria, reduces the net oxygen production by the plants. Therefore the device of suspending both transparent and black bottles is used. The difference in oxygen content between the cultures of raw water in the transparent and in the black bottles after exposure (assuming that respiration by the plants, animals and bacteria is the same both in light and in the dark) is the oxygen produced by photosynthesis (the *gross production*).

Since the volume of oxygen produced nearly equals the volume of CO_2 fixed by the plants, for every $22 \cdot 4$ cm.^{3} of oxygen pro- ★ duced in photosynthesis (difference between transparent and black bottles), 12 mg. of carbon has been fixed. This quantity is the gross production of organic carbon which includes the quantity which has been simultaneously lost by respiration by the plants—a very small fraction during rapid growth.

The most complete series of observations using diatom cultures in transparent and black bottles suspended at various depths in the sea have been made by Jenkin (1937). The bottles were suspended for varying periods up to 24 hours, and measurements of the light of several wavebands were made at intervals. These allowed the light energy falling on the bottles to be calculated.

Fig. 41 presents the results of a typical experiment, and shows markedly less production in depths less than 5 m. on a sunny day. The several experiments show that for light fluxes over a 24-hour period having an average energy of less than about $0 \cdot 03$ g.cal. per cm.^{2} per min. (5 kilolux), production is proportional to light energy. (Under these conditions the utilization of light energy falling upon diatom cells amounted to about 7%.) At average intensities between about $0 \cdot 03$ and $0 \cdot 06$ cal. per cm.^{2} per min. the cells became light saturated, when a considerable further increase in light energy caused little change in oxygen production. At intensities greater than $0 \cdot 16$ g.cal. per cm.^{2} per min. rearrangement or contraction of the chloroplasts in the cells took place.

The depth of the photosynthetic zone above which photosynthesis exceeded respiration in these summer observations in clear offshore water (Secchi disk visible at 12–14 m. depth) was about 45 m. for the two species of diatoms used. The average light energy arriving at the bottom of the zone (the compensation point) was $0 \cdot 002$ g.cal. per cm.^{2} per min. A similar value had been found by Pettersson, using a mixed community of phytoplankton species.

In the winter in latitude $52°$ N. the light falling on the sea daily is about one-ninth of the quantity falling in the summer.

In clear water the quantity penetrating below the 16 m. level in summer is comparable to the quantity entering the water in ★ winter. This indicates that the depth of the compensation point in winter is some 29 m. in clear offshore water in this latitude.

A relation between depth and the CO_2 fixed by phytoplankton, similar to that shown in fig. 41, has been found by Nielsen (1952).

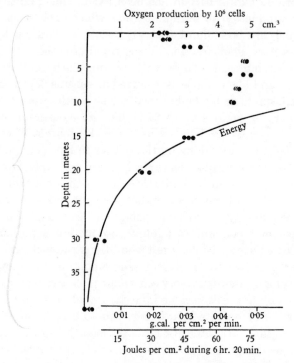

FIG. 41. Oxygen evolved by 10^6 cells of *Conscinodiscus concinnus* while suspended at different depths below the surface for 6 h. 20 m. is shown by dots. The light energy is shown by a solid line. (After Jenkin, 1937.)

A quantity of raw water containing phytoplankton was filled into bottles after the addition of a minute quantity of carbonate containing ^{14}C. After a period of immersion at different depths, the contents of the bottles were filtered, the cells washed and the ^{14}C fixed was estimated.

Using this tracer technique, maximum carbon fixation by phytoplankton was found at about 10 m. depth in the Benguela

Current, an area of upwelling water and extremely rich plankton (fig. 42).

Experiments by Ryther and Vaccaro (1954), using this very sensitive technique have shown that, after short periods of illumination, the increase in ^{14}C in the cells provides a direct measure of the CO_2 photosynthesized. This approximated to the gross oxygen production (difference between transparent and black bottles). The two techniques gave closely comparable values with rapidly dividing cells.

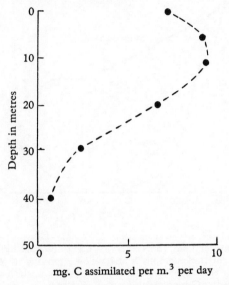

FIG. 42. Production of organic carbon by the plants in water collected at 10 m. depth when exposed at different depths in the Benguela Current. (After Nielsen, 1952.)

With periods of illumination longer than 26 hours the rate of carbon fixation estimated by radio-carbon was less than the rate of photosynthesis as estimated by gross oxygen production. This is attributed to the loss of ^{14}C by respiratory oxidation of the newly formed tissue containing radio-carbon, which had built up to a material fraction of the total tissue after 26 hours' illumination.   ★

Quality of light and growth of phytoplankton. Almost all the biomass of phytoplankton consists of brown diatoms and flagellates. Green, red and blue-green species are exceptional, although they may attain temporary abundance in particular areas. The pigments of the brown algae consist of chlorophylls a and c, with fucoxanthin and other carotenoid pigments. These occur in the granae of the chloroplasts, perhaps in colloidal state or linked to protein. (On denaturating the proteins by heat or other means the brown algae turn green.) When the pigments

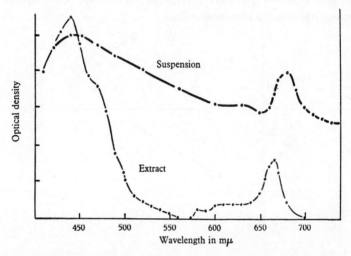

FIG. 43. Light absorbed by a suspension of *Nitzschia closterum* (*Phaeodactylum*) from a light path of 1 cm. (upper curve). Light absorbed by an extract (in methyl alcohol) of the chlorophylls and yellow pigments, the extract having been made up to the original volume of suspension from which it was obtained (lower curve). (From observations by P. E. Potter.)

are dissolved and extracted with a solvent they yield a yellow-green solution. This absorbs very little light of wavelengths between 500 and 630 mμ, that is, green, yellow and yellow-orange light (fig. 43).

Yet, when illuminated with yellow-green or yellow light, diatoms make similar growth as when illuminated with the same light energy of blue or red light, which the extract absorbs readily (Stanbury, 1931).

The rate of photosynthesis by a fresh-water diatom, *Navicula minima*, had been measured by Tanada (1951) with light of various wavebands. The light energy absorbed by the cells during the course of each experiment was measured in terms of quanta of light energy. A quantum of light energy equals $\frac{47 \cdot 5 \times 10^{-17}}{\text{Å.}}$ g.cal., where Å. is the wavelength in Ångstroms.

From the quanta of light energy absorbed and the quantities of oxygen evolved the *quantum yields* (number of molecules of CO_2 reduced, by photosynthesis, per quantum of light energy) were calculated. The action spectrum (fig. 44) shows the mean values found in the numerous experiments. They indicate that

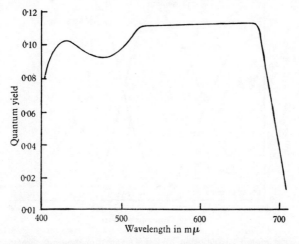

FIG. 44. Action spectrum, being the quantum yield, or number of molecules of CO_2 reduced by photosynthesis per quantum of light energy, by *Navicula minima*, as a function of *wavelength*. (After Tanada, 1951.)

the rate of photosynthesis, at the low light intensities used, is proportional to the light energy divided by the wavelength throughout the range 520–680 mμ. At around and below 480 mμ a smaller proportion of the light absorbed is used for photosynthesis.

The pigments from this diatom were extracted and separated. Fig. 45 shows the percentage absorption of light of various wavelengths by chlorophylls, by fucoxanthin and by 'other carotenoids'. It suggests that the dip at 480 mμ in the action spectrum is due to non-utilization of light absorbed by 'other carotenoids'.

Additional information- is supplied by experiments with brown seaweeds, notably those by Haxo and Blinks in 1950. In effect, they illuminated the thin thallus of a brown weed with light beams of narrow wavelengths, each beam having the same light energy in terms of quanta, and measured the rates of oxygen production. These rates when plotted against wavelength produced a curve (the action spectrum) almost identical

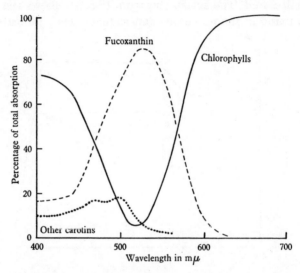

FIG. 45. Estimated distribution of light absorption among pigment groups in live cells of *Navicula minima* as a function of wavelength. (After Tanada, 1951.)

in form to the absorption spectrum of the thallus. This absorption spectrum of the thallus is quite different to the absorption spectrum of the extracted pigments (fig. 46).

For light at intensities below light saturation, it appears that effective use of different wavelengths between 400 and 700 mμ varies with its absorption by chlorophylls and by fucoxanthin *in their natural state*. In their natural state they are, in part at least, colloidal and perhaps in some form of combination with the proteins, fats and waxes of the chloroplast granae.

The light energy absorbed by fucoxanthin in intact *Nitzschia* cells appears to be transferred to the chlorophyll present in the cell. The red fluorescent light, of over 680 mμ, emitted by

Nitzschia cells when excited by light of 600 or 470 mμ, was the same per quantum of exciting light energy. In acetone extracts, on the other hand, light absorbed by fucoxanthin did not contribute appreciably to chlorophyll fluorescence (Dutton, Manning and Duggar, 1943).

The reduction of 1 cm³ of carbon dioxide to sugar with the evolution of 1 cm³ of oxygen requires the utilization of 5 g.cal. In Miss Jenkin's experiments (fig. 41) it was possible to relate the light energy falling on the diatom cells to the oxygen produced. An assumption is made that no one cell casts a shadow on another and that all are lying horizontally. Then during the interval of time required for the production of 1 cm³ of oxygen,

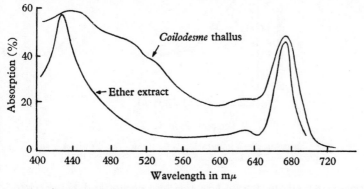

FIG. 46. Absorption of light of different wavelengths by the thallus of *Coilodesme* (upper curve) and by an ether extract containing the pigments. (After Haxo and Blinks, 1950.)

some 72 g.cal. fell upon the cells, when the bottles were at depths below 15 m. Hence about 7 % of the light energy falling on each cell was utilized in photosynthesis.

Near the surface of the photosynthetic zone during bright sunlight, there is a marked inhibition of photosynthesis. This is apparent when the light energy exceeds 0·16 g.cal. per cm² per min., and is accompanied by contraction of the chloroplasts. This may be due to the effect of ultra-violet light. Short wavelength ultra-violet—the actinic rays—are rapidly absorbed by coastal water and completely absorbed by glass. Longer wavelength light, of near ultra-violet, which passes ordinary glass, penetrates the sea to some extent (fig. 40). Experiment suggests that near ultra-violet light may cause partial inhibition of photosynthesis. The growth of *Ditylum brightwelli* was completely inhibited by light from a mercury lamp which had passed through glass, but the diatom grew well in the same light which had ★ passed through a filter which absorbed 'near ultra-violet' light.

TEMPERATURE

The effect of temperature on organic production by *Nitzschia closterium** has been investigated by Barker (1935), who found an increase of about 10 % per degree rise in temperature

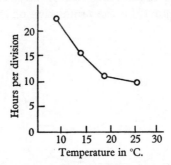

FIG. 47. Change in mean generation time (hours per division) with temperature of *Nitzschia closterium*. (After Spencer, 1954.)

between 15 and 25° C., and rather less above and below that range with a maximum rate at 27° C. Spencer (1954) has found the rate of division of the same species when growing under conditions of light saturation (fig. 47).

In a noteworthy experiment by Wassinck and Kersten (1944) the fresh-water diatom *N. dissipata* was illuminated with yellow light of 589 mμ. With light of low intensities (having light energy below about 0·015 cal. per cm.^{2} per min., comparable to the average light energy below about 20 m. in the English Channel during the summer) temperature had little or no effect upon the production of organic matter (fig. 48).

* There are two marine plants, similar in shape, the diatom *Nitzschia closterium* and the smaller *Phaeodactylum tricornutum*, which has until recently been known as *Nitzschia closterium* forma *minutissima*, and which is also found as triradiate cells. It has characteristics which are not shared by many species of diatoms. It is not always stated in accounts of experiments which of these two plants has been used.

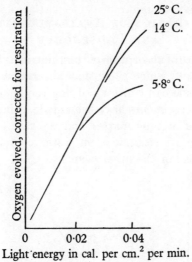

FIG. 48. Oxygen evolved during illumination at different light intensities and temperatures by *Nitzschia dissipata*. (After Wassinck and Kersten, 1944.)

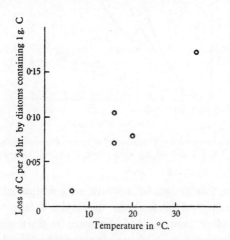

FIG. 49. Respiratory coefficient of *Nitzschia closterium* and of *Coscinodiscus excentricus*. (After Riley, 1949.)

Temperature has a marked effect upon the rate of respiration of diatoms in darkness. Riley *et al.* (1949) have collected the very small amount of available data which can be expressed in terms of percentage loss of organic carbon per 24 hours (fig. 49).

CONCENTRATION OF AVAILABLE NITROGEN
AND PHOSPHORUS

Studies of the rate of absorption of these nutrients have been mentioned (pp. 37–39), which show that absorption and synthesis into organic tissue can continue during the hours of darkness.

Two sets of experiments have been made, concerning the rate of production of organic matter and the rate of cell division, respectively, by *N. closterium*, in water containing low concentrations of one or the other element.

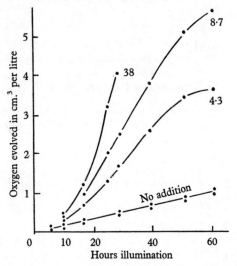

FIG. 50. Oxygen evolved by *Nitzschia closterium* (*Phaeodactylum*) in waters containing 4·3, 8·7 and 38 mg. phosphate-P per m³ (Harvey, 1933).

The diatom was grown in a water with ample nitrate and a limiting quantity of phosphate until the latter was almost exhausted. After keeping the diatoms in darkness for some hours, various additions of phosphate were made and the oxygen evolved was measured at intervals during illumination. The results are shown in fig. 50.

It is of interest that the diatoms resulting from the addition of 4·3 mg. phosphate-P per m³ contained carbon and phosphorus in a ratio of *c.* 100 C:0·4 P, whereas diatoms which are not phosphorus-deficient have a ratio of about 100 C:1·8 P.

Experiments by Ketchum (1939) with the same species show a reduction in rate of cell division when phosphate present in the medium is less than some 17 mg. phosphate-P per m³. One experiment also showed that, with a sufficiency of phosphate, growth was as rapid in a sea water to which 47 mg. nitrate-N per m³ had been added, as in waters with greater additions. (In other experiments the rate of nitrate absorption was reduced when the nitrate content of the medium was below 100–150 mg. nitrate-N per m³.)

There is indirect evidence that the growth rate of phytoplankton *in nature* is reduced when the concentration of these nutrients falls below the threshold values suggested by the above experiments. In an analysis of data collected in the Gulf of Maine, Riley *et al.* (1949) were able to relate the standing crop of phytoplankton present on five occasions during a year with light intensity, temperature and prevalence of zooplankton, provided that an allowance was made for the subthreshold concentrations of phosphate which were present between May and November.

There are several observations which may have a bearing upon the relation of these concentrations to growth rate. The growth of *Phaedodactylum tricornutum* proceeds at a perfectly uniform rate provided that the culture is aerated at a constant temperature and light intensity. If either available nitrogen or phosphorus in the medium is exhausted by the diatoms, their rate of increase in numbers does not immediately fall, but one or more divisions are made before decreased rate becomes apparent (Spencer, 1954).

Munk and Riley (1952) point out that in nature eddy diffusion, both lateral and vertical, will aid the supply of nutrients from very low concentrations in the surrounding water, and also that sinking plant cells will be able to absorb faster than those having ★ the same specific gravity as the surrounding water.

Experiments on the division rate of several species of diatoms in culture have shown that smaller species, which have a greater ratio of surface area to volume, grew more rapidly than larger species when continuously illuminated (Braarud, 1945). It is not surprising that in warm-water areas and in temperate areas during summer, where or when nutrient concentrations are low, the flora tends to be composed of smaller species than in nutrient-rich waters.

Supply of available Iron and Manganese

There is some reason to expect that, in deep ocean areas far from land, lack of essential iron and manganese in available form may reduce the rate of plant growth. This expectation is based on the following premises: (i) practically all the iron and much of the manganese is present in the sea as particles; (ii) particles of ferric hydroxide and of manganese oxide collect

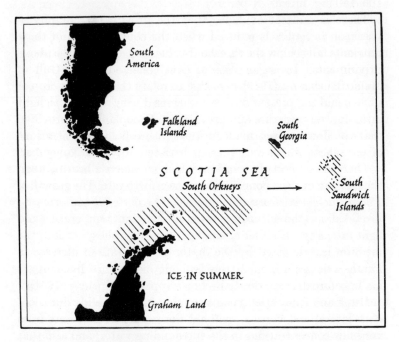

FIG. 51. Areas in the Antarctic which are notably rich in phytoplankton.

on the surface of diatoms and, in consequence of the round of events in the sea, much will be carried downwards; (iii) manganese ions in sea water are readily adsorbed on organic detritus; (iv) aged particles of ferric hydroxide are less readily available to plants than recently formed particles (references to the observations on which these conclusions are based are given on p. 144); (v) the addition of iron and of manganese to sea water collected offshore, which has been enriched with phosphate and nitrate, has often been seen to lead to more rapid growth and larger crops of diatoms or phytoflagellates. This is very indirect evidence, since there may have been sufficient of these elements in the natural water for the relatively small populations which occur in the sea. Nevertheless, it is easy to grow these plants so that they become iron- or manganese-deficient with impaired growth rate, in spite of almost certain contamination of the cultures with these ubiquitous elements.

There is no direct evidence that lack of either of these elements reduces the growth rate of phytoplankton in nature. However, observations by Hart (1942) in the South Atlantic suggest that it is likely. In the Scotia Sea, fed by a current which has washed outlying islands of the Antarctic continent and passed over a submarine ridge, the standing crop is twice as great as elsewhere in the same latitude, and develops earlier. The shaded areas in fig. 51 indicate notably rich plant life.

It would be interesting to investigate the concentration of these elements in available form in waters far from land; but a technique of estimating available iron remains to be evolved, and the existing method for estimating manganese in solution is indirect, being an estimation of its catalytic activity. ★

Vertical Eddy Diffusion and Upwelling

If there were no vertical mixing by eddy diffusion in the oceans, the photosynthetic layer would soon be stripped of plant nutrients, upward movement by molecular diffusion being infinitely slow. Some of the nutrients originally present would be returned by the zooplankton, to be used again by the plants; but soon the zone would be finally stripped. Gardiner's experiments (1937) suggest that a material proportion might be returned by zooplankton while feeding within the layer, to be used again; he collected *Calanus* at sea and added them at once to a vessel of sea water, estimating the increase in dissolved phosphate which soon took place. These early experiments have never been extended.

There are some areas where the upper layer of water moves away as a surface current, usually propelled by wind which may be seasonal. Its place is taken, in part at least, by nutrient-rich

7-2

water from below. This constitutes *upwelling*. The photo-synthetic zone is continuously enriched and organic production is heavy.

Except in shallow and turbulent areas, a thermocline is found, permanent in low latitudes and seasonal in higher latitudes. Over vast areas in low latitudes in the Atlantic it lies near the base of the photosynthetic layer; at 50° N. in the north Atlantic, at some 25 m. in summer; in the shallower water of the English Channel, at 17–18 m., deepening to 23 m. after a summer gale. In high latitudes the thermocline is of short duration and may be limited to a shallow layer.

A result of the thermocline is to hinder replenishment of nutrients from the water below, but not entirely to stop re-plenishment in these upper parts of the photosynthetic zone. Zooplankton rising at night excrete phosphate and ammonium. In addition, there is likely to be some eddy diffusion, since the layer above the thermocline, or *epithalassa*, is free to slide over the water below, and any such movement of one layer over the one below surely increases intermingling.

Thus the upper part of the photosynthetic zone may be cut off to a varying extent from the lower part.

There are other areas where a current runs over an uneven bottom, as over the Grand Banks off Newfoundland, and there-by sets up vertical eddies which extend to the surface. Such eddies have been seen on a perfectly calm day as a ripple on the surface over the Porcupine Bank, which lies off the west coast of Ireland and has 160 m. of water above it. In the deep oceans where the water is only moving very slowly over the bottom, eddy diffusion due to the above cause is unlikely, but may be caused by the internal waves which are found in mid-depths, rising and falling with considerable amplitudes.

All over the oceans wave motion sets up eddy diffusion, the turbulence depending upon the height of the waves, and the depth at which it dies out depending upon the amplitude of the waves or swell. This depth also depends upon the stability of the water column, that is, the increase in density with depth. To mix heavier water below with lighter water above requires much work; eddy diffusion is damped down in any layer where the density is increasing rapidly with depth.

Thus the upward diffusion of nutrients through the base of the photosynthetic zone during any short time interval depends upon (i) their concentration gradient at about this depth; (ii) the degree of eddy diffusion, which is dependent upon wave motion, turbulence arising in the deeper layers and the lateral movement of water layer over water layer on the one hand, and upon the damping effect of the density gradient on the other hand—these all influence the interchange of water below with

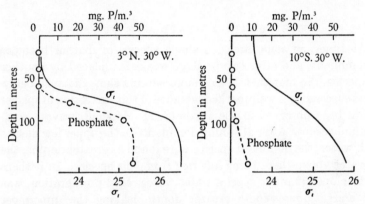

Fig. 52. Phosphate concentration and density *in situ* at two positions in mid-Atlantic, within and south of the equatorial current. (Data from *Discovery* Expedition *Reports*.)

water above; (iii) any upwelling or bodily movement of water to replace water above which has moved away as a current.

Comparing the phosphate gradient, and also the density gradient, in the area of the equatorial currents with that to the south (or north) shows the influence of replacement by upwelling.

Adverse influence of turbulence. In temperate seas, particularly in winter, turbulence carries a proportion of the growing plants downwards below the base of the photosynthetic zone. As a result, the spring increase in phytoplankton may be greatly delayed (Gran, 1932; Hart, 1934, 1942; Riley, 1942). In the English Channel the effect of two or three calm days in January or February on the plant population has been noticed on several occasions. Sverdrup (1953) discusses the effect of the depth of ⋆ the well-mixed top layer of water in relation to the depth of the compensation point and to the standing crop of phytoplankton from March to May at 66° N., 2° E.

A survey of the species distribution and population density of phytoplankton over a wide area of the North Sea and adjacent waters in May has indicated that instability of the water column played a major role in affecting population density. A stable water column, in which vertical eddy motion is damped, permitted greater growth (Braarud, Gaarder and Grontval, 1953).

PHYSIOLOGICAL STATE AND BUOYANCY

At times, in some areas, a single species of diatom becomes dominant, and after two or three weeks is replaced by another species, the first then being found sinking some distance below its successor. A well-marked instance of this succession of species has been observed in the Clyde (S. M. Marshall, private communication). After attaining dominance in the upper few metres of water, the dominant species were found at greater depths, the water above them being left free from this species. In it there developed another species which, after rapid proliferation, was found at increasingly greater depth, leaving the uppermost layers free. Again this uppermost layer became populated with a rapidly growing species and the process was repeated.

Thus in nature the physiological state and the buoyancy of phytoplankton are liable to change.

The cause of these changes is obscure. The reason why a diatom is able to remain buoyant has been investigated by Gross and Zeuthen (1948). From observations of *Ditylum*, which contains a considerable volume of sap, they conclude that when growing freely and remaining in suspension the specific gravity of the sap is 0·0025 less than that of sea water. If growth is checked, the salts in solution in the sap come into equilibrium with the surrounding sea water and the cells sink. They showed that two isotonic sea waters, one of which was deficient in calcium, magnesium and sulphate, had similar differences in specific gravity to those between cell sap and water. This suggests that the mechanism underlying the buoyancy in plankton diatoms consists of 'the maintenance of very low concentrations of divalent ions in the cell sap, the result of a steady expenditure of energy', thereby the cell sap is lighter than the sea water although isotonic with it. Thus, when a succession of species takes place and one species starts to sink, its energy-expending mechanism has ceased to function owing to some cause which does not affect the next species in succession, and in addition its capacity to divide becomes impaired. It begs the question of cause to state that the cells become senile, and only alters the question to consider that the cells may have lost their capacity to form auxospores and continue division.

There is reason to suppose that some species of plants in nature are able to make much more rapid growth, under the same conditions of illumination, in waters from some positions or depths than in others containing similar concentrations of available nitrogen, phosphorus, iron, manganese and silicon. This supposition rests upon experiments of growth in culture (Harvey, 1937; Matudaira, 1939, 1942) which have shown clear-cut differences between growth in different waters similarly enriched.

Whether such differences between waters affect the production of vegetable organic matter by the whole phytoplankton community, or merely affect the species constitution of the community is not apparent. Nevertheless, to consider the growth of these plants in culture is pertinent to the present study.

GROWTH IN CULTURE

A few species of phytoplankton are notably unexacting in their nutritive requirements. Amongst these the very small plant having the appearance of a diatom and known as *Nitzschia closterium* forma *minutissima* or *Phaeodactylum tricornutum* has been much used for experiments on growth rate and environ- ⋆ mental conditions. It has been in unialgal culture for nearly 50 years and has been grown free from bacteria in sea water, or artificial sea water, enriched with nitrogen, phosphorus, iron and manganese. In addition to the original culture isolated by Allen and Nelson (1910) several others have been isolated from cells occurring in different sea areas. (It will be interesting to find whether these different strains have the same physiological requirements, such as optimum temperature and equal growth rates with ammonia, nitrate or nitrite as nitrogen source.) Unlike most, perhaps unlike all, diatoms, this plant produces heavy crops when grown in enriched sea water, without added silicate or silica sol, in hard glass vessels. The plants grown in this manner are very lightly silicified (silica about 1% of dry weight); yet if silicate is added they absorb it rapidly. When this species is grown without aeration, its growth rate starts to slow when the pH of the medium has risen to about 8·6, but growth in a stoppered vessel does not cease until pH 10 is reached.

Besides this plant some other species are strictly autotrophic, persisting through an indefinite number of subcultures and growing through a wide pH range. Strangely enough, these unexacting rapid-growing species are relatively rare in the open sea, considerable populations only being found in some particular, although widely distributed, inshore areas.

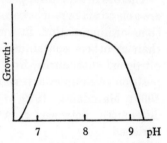

Fig. 53. Effect of pH on growth rate of a mixed culture of marine *Nitzschia* and *Navicula* species. (After Bachrach and Lucciardi, 1932.)

Other species, which are slightly more exacting in their requirements, will behave in the same way, provided the natural sea water which has been enriched happens to be suitable. The diatom *Skeletonema costatum* is such a species and has been kept for many months in persistent culture before dying. However, if it is introduced into a natural water which is unsuitable, the subculture soon dies.

Matudaira (1939, 1942) quotes experiments where waters collected from the surface, enriched and inseminated, produced a heavy crop of dividing cells, whereas in waters collected from a depth of 2, 5 and 10 m. they soon ceased dividing, producing only a small crop of abnormal cells.

Cultures of these slightly exacting species have often lived successfully through a number of subcultures before dying in 'Allen-Miquel sea water' (Allen and Nelson, 1910), which is thought often to give better results than an ordinary enriched sea water. In the preparation of this sea-water culture medium a copious precipitate of iron hydroxide and phosphate is formed, which perhaps carries down with it inhibitory substances which may be in solution in some natural sea waters. The iron and manganese necessary for the plants' growth are supplied by including a trace of this precipitate in the culture, without which the 'Allen-Miquel water' is ineffective.

Yet other species are more exacting and rarely continue for many divisions in sea water enriched with phosphate, nitrate, iron and manganese. The classical experiments by Allen (1914) were the first to show that some marine diatoms were not strictly autotrophic, but required traces of dissolved organic matter to make continued growth. Many exacting species will develop as persistent cultures in *Erdschreiber*, which is natural sea water enriched with nitrate and phosphate to which a boiling water extract of soil has been added. This 'soil extract' provides iron, manganese, a small quantity of silicate and an array of biologically active organic compounds. These include thiamine, biotin, the cobalt-containing vitamin B_{12} and several amino acids. Each has been found to promote the growth of one or other phytoplankton species, and each has been found in lake or pond water (p. 151); a water collected off Halifax, Nova Scotia, has been estimated to contain $0\cdot01\,\mu$g. per litre of B_{12} (cobalamine) by bioassay using *Stichococcus*, a green unicellular alga (R. A. Lewin, 1954).

In both 'Allen-Miquel' and *Erdschreiber* enriched waters, the quantity of silicate in solution does not usually exceed $0\cdot5$–1 mg. Si per litre. If diatoms are grown in it, this is soon used and the diatoms become silica deficient. Then continued growth is dependent upon solution from the glass vessel, which may be very slow from the hard glasses now in general use; growth may even be brought to a standstill and death result if illuminated continuously. The quantities involved merit consideration. Various analyses of diatoms quoted by Vinogradov (1953) show a Si/P ratio of 16 to 50, varying greatly with the species. An analysis of *Skeletonema costatum* collected here showed a ratio of 25. It is not unusual to enrich sea water to the extent of 1 mg. phosphate-P per litre and ten times this quantity of nitrate-N. If this is all to be used in producing diatoms which are not silica deficient, it would be necessary to load the water with

25–50 mg. silicate-Si per litre. This can be done without a pre-cipitate forming at ordinary temperatures by adding a neutral silica sol, which changes to silicate in sea water, but at such high concentrations there is some inhibition of diatom growth.

As growth of an alga proceeds in an enriched sea water, bicar-bonate changes to carbonate, thereby supplying the carbon dioxide used for photosynthesis and the pH of the medium rises until the slow absorption from the air balances the rate of utilization by the algae. As the pH rises above a threshold value growth rate slows. This threshold is about 8·6 for *Nitzschia* (Spencer, 1954) and probably less for many other species.

When grown in stoppered vessels, growth of many species ceased entirely at pH 9·0, when about 8 mg. carbon per litre had been syn-thesized; some less exacting species raised the pH higher—*Skeletonema costatum* to 9·4, *Nitzschia closterium* (*Phaeodactylum*) and *Licmophoia* to 10·0.

It is convenient to consider growth in culture as consisting of two processes, of production of carbohydrate-carbon by photosynthesis and of cell division, which requires many chains of organic syntheses before it can be completed.

When a species requires soil extract or some biologically active organic substance in order to make continued division in a particular sea water (enriched with the inorganic necessities for strictly autotrophic growth) the cells which cease to divide without this organic addition are usually very full of tissue, and are often of abnormal shape. Cessation of growth appears due to cessation of some particular step in one or more of the many series of syntheses, subsequent to photosynthesis, which are necessary for cell division to take place. Similarly, a lesser growth rate (longer mean time of division) at the same tem-perature, light intensity and inorganic nutrient concentration is presumably due to a reduced rate of one or more steps in series of synthesis which finally permit cell division.

There are now numerous examples where one or more steps in a series of syntheses holds back the rate of cell division.

An inability to synthesize B_{12} (cobalamin) or to synthesize it quickly enough for rapid growth is a characteristic of many phytoplankton species. A supply of B_{12} in the water has been found essential for the growth of a number of algal flagellate species in bacteria-free culture; other species required a supply of thiamine or of thiamine in addition to B_{12}; yet other species required one or more biologically active organics in addition, such as biotin and one or more amino acids (Provasoli and Pintner, 1953). R. A. Lewin (1954) notes that the marine *Stichococcus* used in his bioassay was unable to grow freely in enriched sea water except in close association with bacterial colonies.

Experiments have been made in this laboratory with numerous species of marine phytoplankton growing in cultures containing soil extract, and not free from bacteria. When transferred to enriched sea water many species made no, or only limited, growth without the addition of B_{12}, and for some the effect of this addition alone did not replace that of soil extract. The addition of B_{12} was found necessary to produce considerable population densities of *Chaetoceras decipiens* and *Nitzschia seriata*, two diatom species which are at times abundant at the weather-ship station 300 miles west of Ireland (Corlett, 1953), and of *Skeletonema costatum*, [1] ★ an inshore diatom. A natural water, in which this species was dominant, after enrichment and illumination produced few cells without the addition of B_{12}, the addition of which led to dense populations, provided some form of divalent sulphur was also added (in this particular experiment thiourea was added). The addition of a cobalt salt in place of B_{12} was ineffective.

Numerous observations have been made where the addition of organic divalent sulphur compounds or of inorganic sulphide markedly increased the growth of diatoms.

The diatom *Ditylum brightwellii*, grown in *Erdschreiber*, made poor growth when transferred to several sea waters collected during the summer enriched with inorganic nutrients; the cells died or died after starting to make auxospores which did not develop. Their capacity to divide was prolonged in these infertile enriched waters if divalent sulphur was added, in the form of cystine, methionine, glutathione, thiamine or biotin (Harvey, 1939).

TABLE 12. *Growth of* Ditylum brightwellii

Water collected from	No. of cells, including auxospores, present after 5 days' illumination in a north window (initial population 28 cells per cm³)	
	No cystine added	10 mg. per litre cystine added
20 miles offshore:		
0 m.	284	2400
5 m.	53	437
6 miles offshore, 0 m.	238	2400
Near coast, 0 m.	1800	2500

Similar experiments have been made by Matudaira (1942), who found that the addition of sulphide had a marked effect on the growth of *Skeletonema costatum* in both enriched natural sea waters and in artificial sea water.

In another experiment with a surface water, enriched and inseminated with the same diatom species, the effect of adding both sulphide and cystine is seen to be more than additive. This was confirmed, using an artificial sea water.

TABLE 13. *Growth of* Skeletonema costatum

Water enriched with inorganic nutrients after collection from	No. of cells after 60 hours' continuous illumination	
	No sulphide added	With 1 mg. per litre Na₂S, 9H₂O
0 m. (Feb.)	2144 (cells normal)	4626 (cells normal, dividing)
2 m.	429 (cells abnormal, growth ceased)	1526 (cells abnormal, growth ceased)

TABLE 14. *Growth of* Skeletonema costatum

	No. of cells after 60 hours' continuous illumination
Surface water enriched with inorganic nutrients	2658
+ 1 mg. per litre Na₂S, 9H₂O	7258
+ 1 mg. per litre cystine	4294
+ 0·5 mg. per litre Na₂S, 9H₂O + 0·5 mg. per litre cystine	9204

Further experiments with this species of diatom, using thiourea as a source of divalent sulphur, have been made by the writer (unpublished). Transferred from cultures containing soil extract to various offshore sea waters (enriched with nitrate, phosphate, iron, manganese, B_{12} and silicate added in the form of neutral silica sol) the diatom usually grew in curved or even tangled chains, in some waters producing only a quite small crop. With the addition of thiourea the chains were straight and the final crops heavy.

These observations suggested that the deposition of amorphous silica from silicate in solution, to form the frustules of diatoms was in some way linked with their sulphur metabolism.

Meanwhile the same conclusion has been arrived at by I. C. Lewin (1954) using a freshwater pennate diatom. ★

Cells of *Navicula pellicosa* were grown in culture medium until silica-deficient. On adding silicate this was absorbed and laid down as amorphous silica, under aerobic conditions. The uptake was prevented if the silica-deficient cells were washed in water free from sulphate. The washed cells had their ability to take up silicate partially restored by sulphate and completely restored by sodium sulphide, thiosulphate, glutathione, cystine, methionine and also by a mixture of sulphate and ascorbic acid.

The uptake by unwashed deficient cells was prevented by 10^{-3} molar cadmium chloride, which blocks sulphydryl groups but did not affect the diatoms' respiration; if glutathione was added in addition to cadmium chloride the ability of the cells to absorb silicate was restored.

The addition of traces of several organic compounds has been found to increase the growth of various species of diatoms (Levring, 1945; Harvey, 1939). Some of these experiments are difficult to interpret because at the time cobalamin was unknown and could have been added as an impurity. It is synthesized by many species of bacteria, is stable, readily adsorbed and ubiquitous. Moreover, in early experiments the need for manganese and a sufficiency of silicate was not recognized, solution from soft glass culture vessels being relied upon to supply enough of the latter. The addition of soil extract results in no considerable increase in silicate unless the extract is alkaline. The need for organic accessory factors may not be limited to unicellular algae; the growth rate of sporelings of sea weeds is increased by the addition of several biologically active compounds which may conceivably occur in sea water. An account of experiments with sporelings by Henkel (1952) gives references to many such observations.

Distinct from the presence of organic substances in the culture fluid which allow increased rate of division by side-stepping some metabolic synthesis or other, the rate of supply of some necessary microelement in inorganic form can hold back the rate of photosynthesis or division or both.

Such microelements appear to fall into two classes—for example, manganese, which is soluble in sea water but depleted during storage of water in a vessel owing to adsorption on the proliferating bacteria, etc., and iron, which is present as particles or micelles which are utilized by the plants. The utilizable concentrations of some of these can be affected owing to their forming complexes with some organic substances.

An extreme instance is the effect of adding ethylene diamine tetra-acetate, which complexes with many metal ions, is non-toxic, and allows considerable concentration of iron to remain in solution.

★ If added to a growing culture the plants die owing to lack of some microelements in available form.

If added together with considerable quantities of iron, zinc, manganese, molybdenum, cobalt and copper, growth can proceed.

Hutner and Provasoli (1951) found that the following addition to 1 litre of sea water, enriched with nitrate and phosphate, provided a very satisfactory medium for the growth of *Nitzschia closterium*:

500 mg. trisodium salt of ethylene-diamine-tetra-acetate
  6 mg. Fe'''
 30 mg. Zn''
 10 mg. Mn''
 10 mg. Mo as molybdate
  5 mg. Cu''

The iron remains in solution; the concentration of copper would be many times toxic if it were not complexed (chaelated).

Growth in this medium with its hydrogen-ion concentration adjusted to pH 8 has been compared with growths in sea water similarly enriched with nitrate and phosphate to which had been added 200 μg. Fe''' (as citrate) and 20 μg. Mn'' per litre where the iron slowly separates as ferric hydroxide particles, many of which adhere to the diatoms. These additions are at about optimum concentration.

In one comparison, made at a light intensity which provided 'light saturation', the complexed addition caused the growth rate almost to double. It remained at this almost double rate throughout many divisions in several successive subcultures (C. P. Spencer, private communication).

This striking observation suggests that the rate of absorption of microelements may have acted as a bottleneck to production of organic matter by limiting the rate at which cell division took place.

It is noticeable, when growing diatoms in culture, that their behaviour and growth in numbers depend to a large extent upon their previous history, whether they have been making rapid or slow growth before being introduced into the culture medium, that is to say upon their 'physiological state'. The following ★ experiment (Harvey, 1939) shows both this and the effect of temperature on growth rate. A culture of *Biddulphia mobiliensis* was divided into two parts. One was then grown in a north window in relatively dim December light. The other portion was grown near a tungsten electric light bulb immersed in a bowl of running water; it received continuous white light of undefined spectral composition, which had an effect upon a selenium barrier cell similar to that given by 18 kilolux of noon daylight. At the end of a week subcultures of each of these cultures were made in sea water enriched to the same extent. Each subculture was divided into ten glass vessels. They were immersed in two water-baths, one kept at 13° C., the other at 18° C., at different distances from an electric bulb which was immersed in each bath. The light which each vessel of culture received was measured. After 72 hours' continuous light the percentage increase in number of cells which had taken place during the 72 hours was determined.

Experiments with several species of fresh-water algae have shown that variations in environment can lead to wide differences in the relative proportions of the metabolic products which accumulate within the organism (Fogg, 1953). The pattern of their metabolism is remarkably flexible, particularly in those species which are least exacting and accept the widest variations in environment.

Thus growth in culture, with ample supply of available nitrogen, phosphorus (and silicate for diatoms) and under conditions of light saturation, is dependent upon:

(i) The hydrogen-ion concentration of the medium (with which is linked the concentration of molecular CO_2 and the proportion of bicarbonate to carbonate), if it rises above a threshold value. This value is higher for some species than others.

(ii) The rate at which the plants can absorb iron and other microelements.

(iii) The presence in the medium of biologically active organic compounds, which, presumably, certain species cannot themselves synthesize as rapidly as they need them for vigorous or optimum growth.

(iv) The temperature and light intensity at which the parent cells of the inoculum had been growing.

TABLE 15. *Growth of* Biddulphia mobiliensis

	Temp. (° C.)	Percentage increase in 72 hours at				
		28,000 'lux'	18,000 'lux'	8,000 'lux'	4,100 'lux'	1,400 'lux'
Cells grown previously in dim light	18	7	68	66	106	87
	13	24	16	14	8	2
Cells grown previously in continuous light at 18,000 'lux'	18	171	236	190	123	98
	13	31	70	105	67	36

The experiments of various workers indicate that other factors may play a part, such as the nature of the bacterial flora, some of the attached bacteria acting as commensals; the possible presence of inhibitors in the natural sea water; and inhibitors or growth promoters discharged into the culture fluid when growths become dense. Well-controlled experiments by Rice (1954) provide evidence of substances, which are inhibitory to the same and other species of fresh-water algae, being discharged from dense cultures into the water.

A number of diatom species, some of which behave as exacting species, have been grown successfully by Chu (1949) in an artificial sea water (see below). However, it is not stated whether the species will survive successive subcultures. Any artificial sea water is likely to contain traces of many organic compounds.

A faint pattern is emerging of the environmental conditions which affect cell division, in spite of the species ranging from the strictly autotrophic unexacting to the other extreme, such as *Prymnesium parvum*. If this plant gets the opportunity, it will envelop and digest a smaller plant, then divide, producing one pigmented plant daughter and one non-pigmented animal daughter (M. Parke, private communication).

It seems remarkable that unexacting species such as *Nitzschia closterium* forma *minutissima* have not become dominant in the open sea; it has only been found close to the shore. It is equally odd that marine species of *Chlorella* are of rare occurrence in the open sea, in spite of their being not only unexacting, but also of their not being digested by copepods.

Artificial sea water (*Chu*)

	Parts per million		Parts per million
NaCl	23,477	$Na_2SiO_3.9H_2O$	10
$MgCl_2$	4,981	$Al_2(SO_4)_3$	3
Na_2SO_4	3,917	$BaCl_2.2H_2O$	0·09
$CaCl_2$	1,102	$RbHCO_3$	0·34
KCl	664	$MnCl_2$	0·2
$NaHCO_3$	192	$LiNO_3$	1
KBr	96	$FeC_6H_5O_7.3H_2O$ (citrate)	0·54
H_3BO_3	26	$CuSO_4.5H_2O$	0·04
$SrCl_2$	24	KI	0·06
NaF	3	As_2O_3	0·03
$NaNO_3$	50	$ZnSO_4.7H_2O$	0·01
Na_2HPO_4	5		

Until explanations of apparent anomalies such as these can be advanced, our knowledge of the growth requirements of phytoplankton organisms remains very incomplete.

CHAPTER VI

INTERRELATIONS BETWEEN PLANT PRODUCTION AND ANIMAL COMMUNITIES

THE foregoing chapter illustrates how the production of vegetable organic matter below unit area of the sea is influenced and determined by physical factors, factors which embrace the return of plant nutrients to the photosynthetic layer.

Fig. 54 illustrates how the nutrient supply influences the plankton population as a whole (zooplankton and phytoplankton) in the South Atlantic.

Some of the factors, which influence the biomass and kind of animals which the production of organic matter is able to maintain, deserve to be mentioned.

From fig. 13 it is obvious that this quantity is in the first place very dependent upon the *proportion of vegetable organic matter which is destroyed by oxidation*, with or without the participation of bacteria. The proportion is surely variable. An attempt to assess it leads to the suspicion that a rather large proportion of the vegetable organic matter goes to maintain the animal population in an area overlying the continental shelf (p. 31). Here and in many such areas there is a considerable bottom-living fauna, and this fauna makes good use of vegetable and animal detritus, providing, in turn, food for demersal fish.

In nature an equilibrium between the standing crop of plants, herbivores and carnivores is continually passing in and out of balance. Light supply changes from day to day, and generation follows generation of zooplankton. An excess of carnivores, which can be only temporary, leads to a deficiency of herbivores and an abundance of plant food. While an abundance of plants lasts, individual herbivores eat more and void more partially digested food, from which much organic matter passes into solution. Likewise an excess of herbivores, which again can only be temporary, will lead to an accumulation of nutrients in the photosynthetic zone and an abundance of plant food at a sub-

115

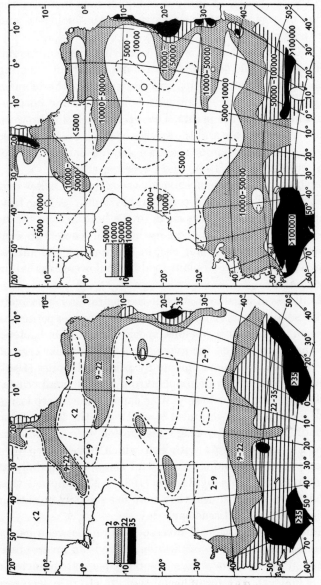

FIG. 54. Distribution of phosphate-P in mg per m³. and of the number of plankton organisms per litre in the upper 50 m. layer in the South Atlantic. (After Hentschel and Wattenberg, 1930.)

8-2

sequent date. The extent and frequency with which the *balance of life* in the sea passes in and out of equilibrium increases the proportion of organic matter destroyed by non-biological and bacterial respiration.

It is well known that in waters overlying the continental shelf the *nature of the bottom* plays a notable part in affecting the biomass of animals present, and consequently the biomass (weight of organic matter) of animals, both pelagic and bottom-living, maintained by the plants. The intimate relation between the nature of the bottom and the species whose larvae will settle upon it and metamorphose is shown by the experiments of Thorsen (1946), Wilson (1948) and Jägersten (1940).

In deep ocean waters the herbivorous zooplankton below unit area is distributed through a considerable depth. A long time elapses before organic detritus sinks to the bottom, and meanwhile it is subject to destruction by bacteria. These two considerations suggest that better use, in maintaining the animal community, can be made of plant food in shallower seas than in deep ocean areas.

The daily loss of organic tissue due to respiration is relatively greater for small animals than for large ones. Thus, experiments by Johnson (1935) with various species of marine bacteria indicate that they lose on the average 15–30 % of their organic matter daily at 12° C.; experiments by Riley and Gorgy (1948) with a mixed community of zooplankton, mostly small crustaceans, showed that immediately after capture they were losing some 12 % daily at 25° C. and 4 % at 12° C., the latter value corresponding with the average respiration rate of *Calanus finmarchicus*. Experiments with fish which have grown to a weight of about 10 g. indicate a loss of some $1-1\frac{1}{4}$ % daily at 12° C., and experiments by Dawes (1930, 1931) showed that small plaice lost weight unless they were fed about $1\frac{1}{4}$ % of their own weight of *Mytilus* tissue each day.

In addition to this general inverse relation between *age or size of animals and losses by respiration*, there is an inverse relation between *age and growth rate* (the daily percentage increase in organic matter). It appears usual that growth rate decreases more rapidly with increasing size than respiration rate decreases. This is shown for *Mytilus* in fig. 55. In consequence, a greater proportion of food assimilated by young animals is built into new tissue than by old animals. Hence the same rate of plant production may permit a greater biomass of a stable community consisting mostly of aged, larger, slow-growing, slow-respiring animals than of one mainly composed of small quick-growing

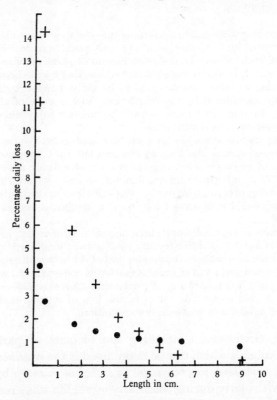

Fig. 55. Relation between percentage daily loss of organic matter due to respiration (circles), percentage daily increase in organic matter due to growth (crosses) and length of *Mytilus edulis*. The smallest individuals, 0·09–0·26 mm., were veligers. (After Jorgensen, 1952.)

animals. The latter fauna, however, may synthesize more animal tissue yearly, the rate of turn-over of living tissue in the animals being greater.

Most marine animals shed great numbers of larvae, whose growth (percentage increase in organic matter daily) is most rapid in their early stages. Most are eaten when young; no more than two survivors out of many need to survive starvation or being eaten in order to replace their parents.

In consequence of the many small rapidly growing herbivores being eaten by larger carnivores, the balance of life within the marine fauna tends towards a maximum ratio of carnivore to herbivore biomass.

It is interesting to speculate concerning the efficiency with which plant food is converted into animal tissue. It is only possible to deal with the hypothetical state where the daily production of plant organic matter remains constant, is wholly digested and assimilated by the herbivores, whose biomass remains constant due to its daily increase by growth being wholly assimilated by the carnivores. Such a state of affairs is never likely to occur in nature, where the animals enjoy alternating periods of overeating and starvation.

Such data as are available for metabolic and growth rates at the average temperature of the English Channel (10–12° C.) suggest that:

(i) 100 g. of phytoplankton organic matter, if entirely assimilated,
★ could yield 70 g. of herbivorous zooplankton organic matter. This 70 g. could yield 4–7 g. of organic matter in pelagic fish feeding on zooplankton, which in turn could yield some 0·3 g. in other carnivores feeding on the pelagic fish.

(ii) Likewise 100 g. of phytoplankton organic matter could yield some 11 g. organic matter as filter-feeding, well-grown, long-lived bivalves and worms such as constitute the greater part of the bottom-living fauna. These 11 g. could yield 1 g. of organic matter as demersal fish, which in turn could yield less than 0·1 g. of predators. These speculative values are maximal. No account is taken of the loss of organic matter by solution and subsequent bacterial decomposition.

Another circumstance affecting the biomass of particular species of animals which can be maintained is the *readiness with which food organisms can be caught in sufficient quantity* by such species, particularly during their early larval life when respiring and growing rapidly and needing much food. If, at this period of their life history, the population density of suitable food organisms falls below a threshold value many larvae die of starvation; they contain relatively small food reserves. Later in life, when food reserves are more abundant and respiratory losses less, most marine animals can survive long periods of starvation.

An illustrative example is seen in the survival of larval cod on the north Norwegian breeding grounds, where they feed upon small copepods, mostly the early stages of *Calanus fin-*

marchicus. During years when there are some ten to twenty copepods of suitable size per litre at the time when the larvae have absorbed their yolk sac, good survival of the larvae has resulted (Wiborg, 1948). Passing back a step in the food chain, *Calanus* have been found to lay large numbers of eggs if phytoplankton is plentiful, but only a few if they are starved (Marshall and Orr, 1952).

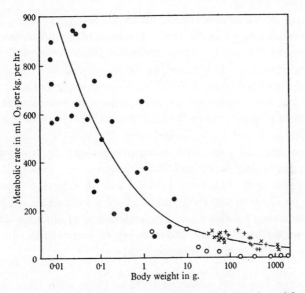

FIG. 56. Relation between metabolic rate and body weight of fish. (After Zeuthen, 1947.)

The part played by temperature is complex. A rise in temperature increases the rate of photosynthesis, but over wide areas of the seas this rate is limited by other factors such as nutrient or light supply. In winter, when nutrients are ample, and light and turbulence control production, there is reason to suppose that temperature may have little effect on the rate of photosynthesis in that part of the photosynthetic zone where light saturation is not attained. On the other hand, temperature has a marked effect on loss by respiration of the plants.

The effect of higher temperatures on the animal community is in general to raise losses by respiration and to increase growth rate. Most animals increase their rate of respiration by about 10 % for 1° C. rise in temperature within a rather wide range. However, some species adapt their metabolic rate after living in a warmer environment; Fox and Wingfield (1937) give instances where the oxygen consumption by some species is less at the same temperature for individuals which have been collected from a warmer environment than for individuals collected from colder waters.

A factor which affects animal life in the sea has recently been discovered. The eggs of *Echinus* have been observed to develop into healthy larvae in water collected from some positions, whereas, in waters from other positions, stunted and unhealthy larvae develop (Wilson, 1951). The filter-feeding behaviour of oysters has been observed to vary widely in waters collected from different positions in the Caribbean. The effect on feeding behaviour is found to be linked with the concentration of an unknown organic substance naturally present in the waters (Collier, Ray, Magnitzky and Bell, 1953).

No simple and ubiquitous relation can be expected between organic production and the biomass of animals maintained by it. This conclusion follows from a consideration of such factors as have been outlined—some obvious, others conjectural.

Yet a general over-all relation exists.

A change has been observed to take place with time in the plant production and in the animals maintained in a particular sea area.

From general observations in waters off Plymouth there is little doubt that the quantity of planktonic and probably bottom-living organisms has decreased during the past 30 years. This opinion is not founded on measurement, but it is reinforced by the results of sampling and measuring particular populations
★ and conditions.

Since 1924 comparable hauls have been made at weekly intervals with a stramin net, having meshes with openings around 1 × 1 mm., each haul filtering roughly 4000 m³ of water. During 1924–30, consistently heavy catches were made, often containing considerable quantities of late-stage copepods and euphausians. During subsequent years, particularly recently, catches have been very sparse indeed. It was soon observed by Russell (1935) that a relation existed between the size of the catches and the phosphate in the water at the beginning of each year.

The average numbers of young fish (excluding clupeids) caught between June and October are plotted against the concentration of phosphate in the water at the beginning of each year in fig. 57. These numbers relate to fish spawned in summer, after the spring outburst of phytoplankton, and after the peak

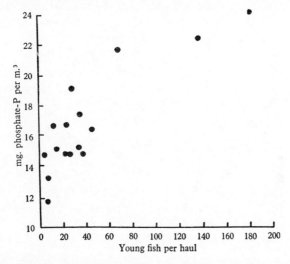

FIG. 57. Relation between the number of young fish caught per haul and the phosphate in the water occupying the area during the previous winter.

population of planktonic crustaceans which immediately follows the spring outburst. These young fish lived their early life during a season when food was not at its greatest seasonal abundance.

No comparison can yet be made of the biomass of animals maintained below unit area in the shallower waters over the continental shelf and of animals in the waters of deep oceans.

In the Antarctic and in areas where upwelling occurs, heavy catches of zooplanton are made and fish are abundant. In other extensive areas, where zooplankton may migrate to great depths, poor catches are usual, notably in the Sargasso. Yet quantitative hauls in the Sargasso, between a depth of 400 m. and the surface, yielded 1–2 g. dry weight of zooplankton

organic matter below a square metre, which was retained by a no. 10 silk net towed at about 1 m. per second (Riley and Gorgy, 1948). This is about half as much as the average maintained below a square metre in the English Channel. Furthermore, the frequency with which 'scattering layers' are found in deep ocean waters suggests a greater animal population than hitherto
★ expected.

PART II

THE CHEMISTRY OF SEA WATER

CHAPTER VII

SALINITY, CHLORINITY, SPECIFIC GRAVITY, OSMOTIC PRESSURE, CONDUCTIVITY

SALINITY AND CHLORINITY

THE exact estimation of the salt content of sea water, by the direct method of drying and weighing, presents difficulties. The sea salts are tenacious of moisture; at temperatures necessary to drive off the last traces, bicarbonates and carbonates are decomposed and bromine with some chlorine is set free and lost. On the other hand, the chloride + bromide content can be measured exactly, since they are completely precipitated as the silver salts. Hence two conventions have arisen, embodied in the terms *salinity* or S‰ and *chlorinity* or Cl‰.

Knudsen (1902) defined the *salinity* of a water as the weight in grams *in vacuo* of solids which can be obtained from water weighing 1 kg. *in vacuo*, when the solids have been dried to constant weight at 480° C., the quantity of chloride and bromide lost being allowed for by adding a weight of chlorine equivalent to the loss of the two halides during the drying.

Thus the salinity equals the weight of the total salt per kilo of water, less most of the weight of the bicarbonate and carbonate ions and less the difference between the bromine and its equivalent of chlorine. An ocean water contains slightly more salts than its salinity value.

The halogens precipitated by a silver salt can be estimated with precision, the precipitate consisting of silver chloride and bromide with an insignificant trace of iodide. This brought into use the term *chlorinity*, by which was understood the mass of chlorine equivalent to the mass of halogens contained in 1 kg. of sea water. This definition implies a knowledge of the exact atomic weights of chlorine and silver. Since the adopted atomic weights have changed from time to time, chlorinity has been redefined (Jacobsen and Knudsen, 1940) in terms of the weight of silver precipitated:

$$Cl‰ = 0.3285234 \, Ag.$$

The relation between chlorinity and salinity has been investigated by Knudsen (1902) for a number of sea-water samples covering a wide range of salinity, those of lower salinity being collected from the Baltic where incoming ocean water is diluted with river water containing small quantities of salts in solution. A straight line relation was found, where

★
$$S\% = 0{\cdot}030 + 1{\cdot}8050\ Cl\%.$$

This relation forms the basis of the tables in general use linking chlorinity with salinity and density.

When an ocean water is diluted with distilled water or ice-melt, the resulting low salinity waters will not have exactly the same density, conductivity and composition as natural Baltic waters of equal chlorinity, from which the above relation was derived.

Present knowledge of the circulation of the water masses occupying the oceans, and much of our knowledge of currents in the upper layers, rests upon small differences in salinity and very small differences in density of the water *in situ*. In practice both are calculated from the chlorinity.

Differences in chlorinity between two samples of water can be estimated with considerable accuracy by titration with silver nitrate, using chromate as indicator and following the customary technique. It is usual to titrate a 'standard sea water', whose chlorinity has been determined by both gravimetric and volume-weight analyses, and so obtain the difference in chlorinity between the unknown and the standard (Jacobsen and Knudsen, 1940). Sealed tubes of such 'standard water' may be obtained from the Laboratoire Hydrographique, Copenhagen; by this arrangement workers in different countries have the same basis for comparison.

Where water masses having very similar salinity need to be distinguished, or where the specific gravity *in situ* needs to be found in order to calculate the direction and velocity of currents, the errors in titrations need to be kept within about 1 part in 3500. This involves using a burette and pipette designed for the purpose, and taking many necessary precautions, in order to extend the precision of volumetric analysis to its utmost limit. Particulars of the technique are given by Thompson (1948), and the necessary tables for conversion to chlorinity, salinity and specific gravity at any temperature by Knudsen (1901) in his *Hydrographic Tables*. A very useful account of the procedure has been published by Matthews (1923).

A solution of chromate is most frequently used as indicator, and more recently a solution containing 0·005–0·05 % of the sodium salt of fluorescein in a 1 % starch solution. This was introduced by Miyake (1939) and has been found by several observers to give a sharper end-point. Even greater precision can be obtained by titrating to a potentiometric end-point (Bather and Riley, 1953).

TABLE 16. *Salinity corrections*

Salinity, S ‰ found	Correction to be applied	Salinity, S ‰ found	Correction to be applied
40	−0·15	22	+0·22
38	−0·08	20	+0·23
36	−0·03	18	+0·23
34	+0·03	16	+0·23
32	+0·07	14	+0·20
30	+0·11	12	+0·19
28	+0·15	10	+0·16
26	+0·17	8	+0·15
24	+0·20		

In order to estimate chlorinity in very small volumes of sea water or body fluids, Keys (1931) developed a technique using a temperature-compensated syringe pipette, which delivers with an accuracy of about 1 part in 15,000, in a rapid and accurate method of the Volhard principle.

A method of estimating salinity from electric conductivity measurements has been in constant use in the Ice Patrol vessels of the U.S. Coastguard Service (Wenner, Smith and Soule, 1930), and a method based on measurement of the refractive index of the water has been used to a limited extent. Many methods based on density determinations have been devised, but do not meet the requirement of use on board ship if great accuracy is required.

For many purposes it is sufficient to titrate 10 cm³ of sea water with a solution containing 27·25 g. of silver nitrate per litre from an ordinary burette. The volume, in cubic centimetres, of the silver nitrate required will roughly equal the salinity of the sample. Actually they are not quite in the same proportion, since 10 cm³ of a more dilute sea water than the 'normal' water will not weigh so much. To allow for this it is necessary to apply the small correction given in table 16.

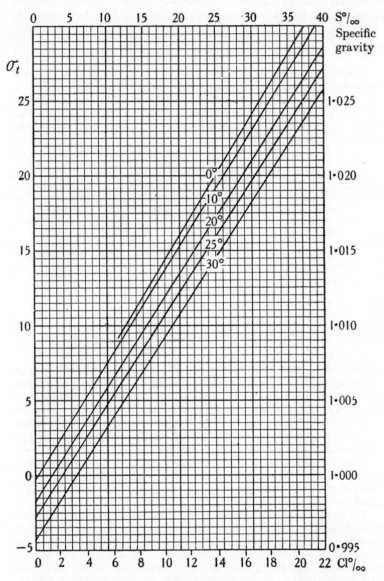

FIG. 58. Relation between salinity, chlorinity and specific gravity
at 0, 10, 20, 25 and 30° C.

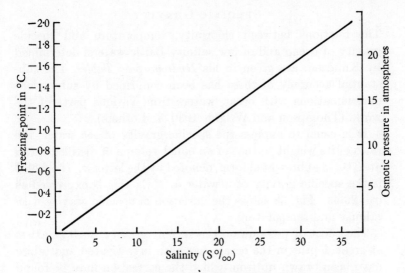

Fig. 59. Relation between salinity, freezing-point and osmotic pressure of sea water.

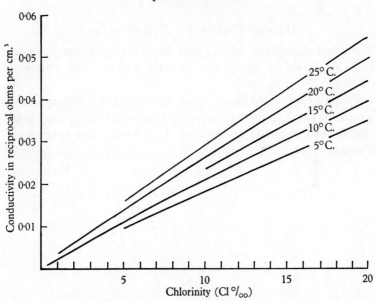

Fig. 60. Relation between chlorinity and electrical conductivity of sea waters. (Data from Thomas, Thompson and Utterback, 1934.)

SPECIFIC GRAVITY

The relations between chlorinity, temperature and specific gravity of ocean and of low salinity Baltic waters, determined by Knudsen, are given in his *Hydrographic Tables*. The substantial accuracy of these has been confirmed by subsequent determinations with ocean waters from various parts of the world (Thompson and Wright, 1930, and others).

It is usual to express the specific gravity of sea water, the ratio of its weight to that of an equal volume of distilled water at 4° C., in abbreviated form, denoted by the letter σ. If 1·02653 is the specific gravity of a water at $t°$ C., this is expressed as $\sigma_t = 26·53$. Fig. 58 shows the variation of specific gravity with salinity and temperature.

The effect of pressure on the specific gravity, such as exists at great depths in the ocean, has been investigated and tables have been drawn up from which the correction may be found (Ekman, 1910).

OTHER PHYSICAL PROPERTIES

The relation between salinity and osmotic pressure or freezing-point is shown in fig. 59 and between salinity and electrical conductivity in fig. 60.

Viscosity relations have been tabulated by Miyake and Koizumi (1948), vapour pressure by Miyake (1952), by Robinson (1954), and refractive index by Utterback, Thompson and ★ Thomas (1934).

CHAPTER VIII

THE MAJOR CONSTITUENTS

THE sea salts of ocean water, away from any considerable dilution by land drainage, are of almost constant composition. Dittmar's analyses of seventy-seven samples of water collected by H.M.S. *Challenger* in 1873–6 from all the oceans showed this constancy. Subsequent analyses, in greater detail, bear out Dittmar's very accurate data. These have been recalculated in terms of the atomic weights of 1939 by Lyman and Fleming (1940).

A few of the minor constituents, such as nitrogen compounds and phosphates, undergo considerable changes since they are utilized by plants. Small changes in the calcium content are brought about in the same way by animals and plants, and also by solution of calcium carbonate from calcareous bottom deposits at great depths.

When ocean water is diluted with land drainage, the percentage composition of the contained salts is altered, since river water contains more sulphate than chloride and more calcium than magnesium than sodium. This change is reflected in Knudsen's formula relating the salt content to the chlorinity, where

$$S\% = 0.030 + 1.8050 \, Cl\%.$$ ★

The following table gives the percentage composition of salts in sea water, which has not been materially diluted by land drainage. It is compiled from the recalculated data of Dittmar, and various subsequent analyses which show the relation of one or other constituent to the chlorinity of the water.

Percentage composition of salts in ocean water

Na·	30·4 %	Cl′	55·2 %
Mg··	3·7	SO$_4''$	7·7
Ca··	1·16	Br′	0·19
K·	1·1	H$_3$BO$_3$	0·07
Sr··	0·04	(HCO$_3'$ and CO$_3''$	0·35)

Minor constituents 0·02–0·03.

In sea water the carbon dioxide is mostly in the form of bicarbonate and the boric acid as undissociated molecules (p. 153). The more recent data concerning these major constituents are given in the following account.

OCCURRENCE OF MAJOR CONSTITUENTS

Sodium. In the early analyses of sea water, the sodium content was found by difference. Recently the ratio of sodium to chlorinity has been found directly by precipitation with zinc uranyl acetate; Webb (1938) obtained a value of $0.5549 + 0.001$ for the ratio, while Robinson and Knapman (1941) found 0.5549 for waters of the North Pacific and a slightly greater ratio for inshore waters.

Magnesium. Dittmar's analyses, recalculated by Lyman and Fleming (1940), give the ratio of magnesium to chlorinity as 0.06801. Thompson and Wright (1930) cite a number of analyses by other observers, and themselves found a mean ratio of 0.06694.

Matthews and Ellis (1928), by precipitating the magnesium with 8-hydroxyquinoline, found a mean ratio of 0.06785 in the eastern Mediterranean and 0.06814 in the Gulf of Aden.

Miyake (1939) finds ratios of 0.0669 and 0.0676 for the northeast and west Pacific respectively.

Wattenberg and Timmermann (1937) find the solubility product of both magnesium carbonate and hydroxide to be greater in sea water than in fresh water,

$$K'_{Mg(OH)_2} = 5 \times 10^{-11}, \quad K'_{MgCO_3} = 3 \times 10^{-4}$$

for ocean water.

When the hydrogen-ion concentration of sea water falls and the pH rises above $c.$ 9, magnesium hydroxide separates as a precipitate with calcium carbonate.

Calcium. In waters of the open ocean away from the influence of land drainage, the calcium content of the water bears an almost direct relation to the chlorinity; very slight divergences are caused by organisms utilizing calcium in the upper layers and by solution from bottom deposits as a result of pressure at great depths.

The relation of calcium to chlorinity found by various observers is shown in the following table, in which allowance has been made for the strontium included in the estimates for calcium (Webb, 1938):

	Ca/Cl
Dittmar (1884) recalculated 1938 atomic weights (Lyman and Fleming, 1940)	0·02095
Thompson and Wright (1930)	0·02090
Thompson and Wright, average of estimates by other observers	0·02177
Kirk and Moberg (1933)	0·02122

When ocean water is diluted with river water, the ratio is affected, since river water usually contains very much more calcium in proportion to chloride than sea water. In low salinity waters of the Baltic (Cl‰ 1–7) several distinct water masses have been distinguished by their Ca/Cl ratio (Gripenberg, 1937).

The quantity of calcium which can exist in solution in sea water is limited by the solubility of $CaCO_3$. If the concentration of carbonate ions is raised, as by blowing a current of alkali-washed air through the water, $CaCO_3$ is precipitated. Considerable supersaturation may be attained before precipitation commences. At high pH values some $Mg(OH)_2$ also separates.

The solubility of $CaCO_3$ is greatly increased by the presence of neutral salts. The 'apparent' solubility product

$$(K'_{CaCO_3} = C_{Ca''} \times C_{CO_3},$$

where $C_{Ca''}$, etc., is the concentration of the ion in gram ions per litre) has been determined (Wattenberg, 1933, 1937) by shaking calcite with sea waters made acid with CO_2, the following values being found at 20° C.:

Salinity ‰	0	5	15	20	25	30	35
$K'_{CaCO_3} \times 10^{-6}$	0·003	0·05	0·10	0·17	0·25	0·35	0·58

The variation with temperature for water at S‰ 32 has been determined in the same manner by Wattenberg and Timmermann (1936):

Temperature (° C.)	0°	10°	20°	25°	30°	35°
$K'_{CaCO_3} \times 10^{-6}$	0·83	0·74	0·62	0·52	0·44	0·40

The concentration of carbonate ions in a sea water can be calculated from the following formula, derived from the definitions of carbonate alkalinity and the second apparent dissociation coefficient of CO_2 (pp. 162, 167):

$$C_{CO_3} = \frac{\text{Carbonate alkalinity} \times K_2'}{2K_2' \times C_H}.$$

Precipitation of $CaCO_3$ from supersaturated sea water brings about a shift in the CO_2 system, since both free base and total CO_2 in solution are reduced. This causes a rise in the hydrogen-ion concentration. Calculated changes agree well with those observed (Wattenberg, 1933, p. 215).

Revelle (1933) has also investigated the 'apparent' solubility product of $CaCO_3$ in sea water, by finding the quantity of calcium remaining in the water after part had been precipitated. Air freed from CO_2 was bubbled through samples of water at 30° C. for several weeks, washing out some of the CO_2 and causing precipitation of aragonite. At the end of the experiments, the pH, excess base, chlorinity and calcium content of the samples were determined. The titrations for excess base were made to an end-point of pH 4·5, that is, some 3×10^{-4} equivalent of acid were required in excess of the amount required to neutralize the excess base. In order to bring the results into line with Watten-berg's values, the solubility products have been recalculated allowing for this and using Buch's values for K_B' and K_2'.

pH	Cl ‰	C_{Ca} mols per litre $\times 10^{-3}$	Excess base corr. $\times 10^{-3}$	$C_{CO_3''}$ mols per litre $\times 10^{-3}$	K'_{CaCO_3} 30° C. $\times 10^{-6}$
8·98	20·95	10·57	0·81	0·175	1·81
9·12	20·60	10·63	0·92	0·222	2·36
8·83	19·60	10·19	1·01	0·236	2·4

Experimental data from Revelle, R. and Fleming, R., *Proc. 5th Pacific Sci. Congress*, 3, 1933.

These values for the solubility product, obtained by pre-cipitating aragonite, are much higher than those found by Wattenberg by dissolving calcite. Equilibrium is attained very slowly, and may not have been completed. Calcite is stated to be less soluble than aragonite.

Smith (1940) has shaken samples of sea water for long periods at 30° C. with calcareous bottom deposits. CO_2 in small quantity was set free in some of the experiments, presumably due to biological oxidation, causing solution of aragonite from the bottom deposit. In other experiments the calcium content of the water decreased owing to precipitation. The results indicated a solubility product in the neighbourhood of $K'_{CaCO_3} = 1\cdot2 \times 10^{-6}$ for 30° C. Similar values have been calculated by Smith from data by Gee, Greenberg and Moberg (1932).

These latter values indicate that the upper layers in the Caribbean are over 100 % supersaturated, even in areas where deposition of aragonite is proceeding. Wattenberg's values indicate from 300 to 700 % supersaturation in such tropical waters, and marked supersaturation in the upper layers in cold latitudes.

With increasing depth and fall in pH due to pressure (p. 179) the waters become undersaturated, and there is evidence of solution from calcareous bottom deposits taking place at great depths (p. 161).

Potassium. Recalculation of Dittmar's analyses by Lyman and Fleming (1940) show a ratio of potassium to chlorinity equal to 0·02029. An average value of 0·02000 from various published data is given by Thompson and Robinson (1932). Analyses by Webb (1938) give a value of 0·02009 ± 0·00020. Miyake (1939) found a ratio of 0·0191 in Pacific waters.

The small quantity of potassium in sea water in comparison with that of sodium is considered due to its more ready adsorption on particles of detritus and its consequent concentration in bottom deposits (Goldschmidt, 1934; Noll, 1931). Since living organisms obtain and concentrate potassium from the waters, they may help in this process.

Strontium. Analyses by Desgrez and Meunier (1926), since confirmed (Thompson and Robinson, 1932), show a strontium/chlorinity ratio of 0·0007. A similar value has been obtained by Webb (1938), who points out that estimations of calcium in sea water included the strontium. Miyake (1939) gives a value of 0·00075 for the ratio. Smales (1951) finds a ratio of 0·0005. ★

Strontium is found concentrated in many algae, in quantity up to 90 times the concentration in an equal volume of sea water

(Black and Mitchell, 1952; Spooner, 1949). Spooner observed very heavy adsorption of yttrium, the daughter product of radioactive strontium, by red and green algae and by a diatom but not by Phaeophyceae.

Sea water is not saturated with respect to strontium carbonate, the presence of neutral salts in water greatly increasing its solubility (Wattenberg, 1937 a, b, c).

The skeleton of a radiolarian, *Podocanalus*, is stated to be composed of strontium.

Sulphate. Thompson, Johnson and Wirth (1931) have determined the ratio of sulphate to chlorinity in a number of samples collected at varying depths from each of the oceans, including waters from the eastern Mediterranean and the Red Sea. Only very small deviations from the value 0·1395 were found. The low salinity waters of the Baltic were exceptional, having a mean ratio of 0·1414. Previous investigations by Thompson, Lang and Anderson (1927) had shown a variation in the ratio in waters diluted by natural means, and in water from which ice had frozen out (Lewis and Thompson, 1950).

Boron. The boron in sea water exists as boric acid, the greater part being undissociated at the hydrogen-ion concentrations ordinarily met with. Its apparent dissociation constants in salt solutions have been investigated (p. 163). Marine plants are rich in boron, their ash being stated to contain 1 % B_2O_3 (Wattenberg, 1938).

The ratio of boric acid to chlorinity has been determined as 0·00137 by Harding and Moberg (1933) and by Igelsrud, Thompson and Zwicker (1938). Miyake (1939) reports an increase in this ratio with depth in Pacific waters, the value changing from 0·00136 to 0·00162.

Bromine. The bromine/chlorinity ratio in sea waters has been determined as 0·00340 by Thompson and Korpi (1942).

Miyake and Sakurai (1952) describe the use of the chlorinity/bromine ratio in following the distribution of water masses in an estuary.

ARTIFICIAL SEA WATER

It is frequently desired to make an artificial sea water from laboratory reagents. Several formulae have been in use, of which the following, due to Lyman and Fleming (1940), includes all the major constituents and yields a water of $Cl = 19·00\%$ and salinity $34·33$:

$NaCl$	23·477 g.	KCl	0·664 g.	H_3BO_3	0·026 g.
$MgCl_2$	4·981	$NaHCO_3$	0·192	$SrCl_2$	0·024
Na_2SO_4	3·917	KBr	0·096	NaF	0·003
$CaCl_2$	1·102				

$$H_2O \text{ to } 1000 \text{ g.}$$

Allowance has to be made for water of crystallization in any of the salts used. After thorough aeration the pH should lie between 7·9 and 8·3.

CHAPTER IX

THE MINOR CONSTITUENTS

THE concentration of the minor constituents in sea water is frequently and conveniently expressed in terms of microgram-atoms per litre (= mg. at. per m.³). One microgram-atom (μg. at.), equals the atomic weight of the element in micrograms and contains 6×10^{17} atoms.

It is noteworthy that even at exceedingly great dilution of an element the number of its atoms in a very small volume of water is very great. For instance, a sea water containing 0·004 mg. gold per m.³ will contain 2 million atoms per mm.³, or one atom in every 500 μ.³ Hence even at these great dilutions many atoms make contact with any surface, of organism or of detritus, in suspension in the sea.

It is noteworthy that, in general, greater concentration of many micro-elements tend to be found in the Pacific than in the Atlantic. Published analyses (Table 17) are not all of waters from the open oceans.

Many of these elements occurring at great dilution in the sea are found concentrated several hundredfold or several thousand-fold in one or more species of marine animals (Noddack, 1940). Black and Mitchell (1952) have investigated the occurrence of a number of micro-constituent elements in nine species of seaweeds.

It seems probable that concentration by living organisms takes place by one or more of three mechanisms. The simplest is adsorption of ions on the cell-water interface or at interfaces within the cell; this seems likely with gold, which Haber in his numerous analyses found to be more abundant in waters rich in plankton. Absorption of radioactive vanadium from solution in the sea and concentration in localized tissues of two species of tunicates has been observed by Goldberg, McBlair and Taylor (1951).

A material proportion of niobium has been found in one species accompanying the vanadium (Carlisle and Davis, private communication).

There are ions in solution, ammonium, nitrite, nitrate, phosphate, cupric copper, manganese, molybdate, cobalt and zinc which, having passed into a plant cell, are built into organic matter and so passed on to animals. There are other elements which, in the form of colloidal micelles, very readily attach themselves to the mucus surrounding plants and animals; ferric hydroxide micelles are a notable instance.

Occurrence of Minor Constituents

Arsenic. Water from the Pacific was found to contain 15–35 mg. As per m^3 by Gorgy, Rakestraw and Fox (1948), who attributed 50–60 % to arsenite, while Ishibashi et al. (1951) found 3–5 mg. per m^3.

In offshore waters from the English Channel Smales and Pate (1952) found 1·6–5 mg. by a radioactivation method and a similar range by the method used by Gorgy et al. Analyses by the writer, using arsenomolybdate colorimetry, yielded 2·4–3·1 mg. As per m^3. When arsenite was added to English Channel waters at pH 8·2, it oxidized to arsenate, rapidly in some, slowly in others.

This element is concentrated in some animals, as the medusa Cyanea (Noddack and Noddack, 1940), oysters, Pecten and Ficulina (Orton, 1923), also in seaweeds (Black, private communication; Jones, 1922).

Arsenate has been found to take the place of a proportion of the phosphate required for growth by the fresh-water alga Stichococcus flaccidus (Comere, 1909); experiments with a marine diatom failed to show that any phosphate could be replaced by arsenate.

A remarkable observation has been made by Orton (1923) that oysters, living in sea water containing arsenite at a concentration of 4 g. As per m^3, produce arsine in sufficient quantity to be detected by its smell.

Copper. Numerous analyses of sea waters by various observers have been tabulated by Chow and Thompson (1952) which show wide variations in concentration of copper, ranging from about 25 mg. Cu per m^3 in the low-salinity water off the mouth of the Mississippi River to less than 1 mg. Cu per m^3 in other areas.

★   TABLE 17. *Minor constituents in sea water*

	mg. per m³ = μg. per litre	mg. at. per m³ = μg. at. per litre	Reference
Aluminium	0–7	0–0·25	Simons, Monaghan and Taggart (1953)
	5–10	0·2–0·4	Armstrong (private communication)
	160–1,800	6–66	Haendler and Thompson (1939)
	1900	74	Thompson and Robinson (1932)
Antimony	0·2	0·0016	Noddack and Noddack (1940)
Arsenic	See p. 139		
Barium	30–90	0·22–0·66	Von Engelhardt (1936)
Bismuth	0·2	0·001	Noddack (1940)
Cadmium	0·032–0·057	0·0005	Mullin and Riley (1954)
Caesium	2	0·0015	Wattenberg (1938)
Carbon, organic	See p. 149		
Cerium	0·4	0·003	Goldschmidt (1937)
Chromium	1–2·5	0·02–0·05	Black and Mitchell (1952)
Cobalt	0·1	0·002	Noddack (1940)
Copper	See p. 139		
Deuterium	16,000	8,000	Friedman (1953)
Fluorine	1,400	74	Thompson and Taylor (1933)
	1,300	69	Miyake (1939)
Gallium	0·5	0·007	Noddack (1940)
Germanium	Identified	Identified	Bardet (1938)
Gold	0·004	0·00002	Haber (1928)
	0·008	0·00004	Noddack (1940)
Iodine	50	0·2	Reith (1930); Schultz (1930)
Iron	See p. 142		
Lanthanum	0·3	0·0015	Goldschmidt (1937)
Lead	4	0·002	Boury (1938)
	5	—	Noddack (1940)
Lithium	100	14	Thomas and Thompson (1933)
Manganese	See p. 146		
Mercury	0·03	0·00015	Goldschmidt (1937)
Molybdenum	0·3–0·7	0·003–0·007	Ernst and Hoermann (1936)
	2	0·02	Bardet (1938)
	0·4	0·004	Noddack (1940)
	12–16	0·12–0·16	Black and Mitchell (1952)
Nickel	0·1	0·002	Ernst and Hoermann (1936)
	0·5	0·01	Noddack (1940)
	1·5–6	0·03–0·1	Black and Mitchell (1952)

TABLE 17. *Minor constituents in sea water (cont.)*

	mg. per m³ =μg. per litre	mg. at. per m³=μg. at. per litre	Reference
Nitrogen			
as ammonium-N	< 5–50	—	
as nitrite-N	0·1–50	—	
as nitrate-N	1–600	—	
as organic-N	30–250	—	See p. 53
Phosphorus			
as phosphate-P	< 1–60	—	
as organic-P	0–16	—	See p. 37
Radium	0·07–0·58 × 10⁻⁷	—	Rona and Urey (1952)
Rubidium	200	2·3	Goldschmidt (1937)
Scandium	0·04	0·001	Goldschmidt (1937)
Selenium	4	0·05	Wattenberg (1938)
Silicon	See p. 147		
Silver	0·3	0·003	Haber (1928)
	0·15	0·0014	Noddack (1940)
Thorium	0·01–0·001	0·00004–0·000004	Koczy (1949) (by calculation)
Tin	3	0·003	Noddack (1940)
	< 5	—	Black and Mitchell (1952)
Titanium	1	0·05	Griel and Robinson (1952)
	6–9	0·1–0·2	Black and Mitchell (1952)
Vanadium	0·2–0·3	0·004–0·006	Ernst and Hoermann (1936)
	2·4–7	0·05–0·14	Black and Mitchell (1952)
Yttrium	0·3	0·003	Goldschmidt (1937)
Zinc	14	0·2	Noddack (1940)
	< 8	< 0·1	Atkins (1936)
	9–21	0·14–0·3	Black and Mitchell (1952)

Atkins (1953) records a seasonal variation in the surface water content of copper 20 miles offshore in the English Channel, ranging from *c*. 25 mg. per m³ in winter to 1·5 mg. per m³ in early autumn. Kalle and Wattenberg (1938) found that surface waters in mid-Atlantic contained less than 3 mg. Cu per m³.

This element occurs in haemocyanin, the respiratory pigment of many invertebrates, in several enzymes and in chloroplasts. It is essential for plant growth.

The solubility of cupric ions (cuprous are rapidly oxidized to cupric) in sea water is limited by the solubility of the green oxycarbonate, which gradually changes to basic carbonate, which is blue-green. The solubility of the basic carbonate in sea water of pH 8 lies in the region of 180 mg. Cu per m³ and is much greater at pH values below 7. When metallic copper is placed in sea water cuprous ions are discharged; these are rapidly oxidized to cupric and heavy supersaturation results, which may persist long enough for a concentration of 2000 mg. cupric Cu per m³ to be temporarily attained.

The concentration of cupric ions which is poisonous to marine plants and animals varies around 1000 mg. per m³, being different for different species—concentrations which are much above the ultimate saturation value at pH 8. Surfaces, such as copper antifouling paints, from which copper dissolves at the rate of 10–12 μg. per cm² per day prevent the attachment and subsequent growth of marine organisms. If a chaelating agent, such as ethylene-diamine-tetra-acetate, is added, sea water can be heavily loaded with copper without precipitation and with reduction or destruction of its poisonous effect on algae.

If sea water enriched to the extent of 50–100 mg. Cu per m³ is shaken with plankton organisms, marine mud, oil droplets or rosin particles, the copper in solution is markedly reduced. Riley (1939) has studied the copper and plankton concentrations in lake waters and found a relation which follows the 'Freundlich adsorption isotherm'.

The copper content of ten species of seaweeds has been determined by Black and Mitchell (1952), who find 0·7–5·7 mg. per kg. wet weight, indicating a concentration of 100–600 times that in an equal volume of the sea water in which they were living.

Iron. *Forms in which iron occurs.* A variable quantity of iron occurs in sea water; from less than 1 to 50 or 60 mg. Fe per m³ have been found in offshore waters. Part is retained by fine-texture filter-paper and all detectable traces, or almost all, by a membrane filter. Part is soluble in dilute acid and a further quantity is dissolved on heating with acid and an oxidizing agent such as bromine water.

The quantity in true solution is extremely small, owing to the insolubility of ferric hydroxide, while the quantity of ferrous ions which can remain in solution is limited by the oxidation-reduction potential of sea water which has been investigated (Cooper, 1937b). From solubility product and activities of ferrous, ferric and FeOH″ ions, Cooper (1937a) has concluded that sea water, when equilibrium has been attained, can contain no more than 4×10^{-7} and 3×10^{-8} mg. per m.^{3} as ionic iron in true solution at pH 8 and 8·5 respectively. In addition to this there may be traces of iron in solution in forms such as haem compounds set free in the breakdown of organisms. Such stable organic iron compounds have not been found in the sea; most organic compounds containing iron are slowly hydrolysed in sea water.

With regard to the presence of colloidal micelles of the hydroxide in sea water, direct experiment has shown that the rate at which they aggregate falls off materially at the very low concentrations which are likely to occur in the sea. Furthermore, they are afforded a considerable measure of protection against aggregation by the presence of almost equally low concentrations of many organic substances, particularly by large molecules containing many hydroxyl groups.

When aggregation takes place with the formation of flocculi, these sediment more or less rapidly. In the open ocean, beyond the influence of land drainage and where the upper layers are less dense than the water below, sedimentation may leave these upper layers almost iron free if turbulence (vertical mixing) is insufficient to keep them supplied with particles from below. The data obtained by Seiwell support this expectation.

The renewal of colloidal micelles or small aggregates is also likely to proceed through the iron collected by plant organisms being eaten with the plants and partly dissolved during digestion. When voided, the ferric hydroxide formed on mixing with the alkaline sea water will be in proximity to much protective colloid excreted at the same time.

Particles of detritus in ocean water, retained on a fine filter, contain material quantities of iron and a significant quantity of phosphate, some of which may be present as aged unreactive ferric phosphate. Yet if a solution of ferric salt is added to sea water, it hydrolyses, flocculates and sediments, and the pH of the water falls.

With very small additions of ferric salt, or of ferric hydroxide, the concentration of phosphate in solution is reduced; with larger additions the water is stripped of phosphate. If recently precipitated ferric phosphate is added to sea water it hydrolyses, turning brown, and sets free phosphate until the pH of the water falls to c. 7.

Distribution. Estimations of iron in offshore waters have been made, notably by Braarud and Klem (1931) off the Norwegian coast, where 3–21 mg. Fe per m^3 were found, while Thompson and Bremner (1935) found larger quantities ranging from 15 to 50 mg. per m^3 in the Pacific. At one position in the Atlantic Seiwell (1935) found no iron in the upper 40 m. layer by a method which would detect 1–2 mg. per m^3

In the English Channel, Cooper (1935, 1947) found variable quantities by a method which estimated the iron soluble on boiling with dilute acid and bromine water. The quantity varied between 0 and 30 mg. Fe per m^3, with occasional high values in samples from near the sea bottom and in samples of surface water. The latter observation suggests a surface film containing entrained particles or micelles of ferric hydroxide. Waters rich in iron were on the average richer in phosphate.

A seasonal change in the iron of the inshore waters of Puget Sound has been recorded by Thompson and Bremner, and a reduction in quantity after the spring growth of diatoms in the English Channel is recorded by Cooper. Various analyses of these plants yielded more iron than phosphorus, which points to the diatoms collecting all the iron in the water more than once during the course of a year. This implication lends particular interest to the nature of the iron present in the sea, since there is insufficient in true solution as iron ions to supply the plant with anything approaching the large quantities found when they were analysed.

Adsorption and absorption by plants. It has been found that marine plant organisms can utilize insoluble ferric hydroxide, and that colloidal micelles or larger aggregates are readily adsorbed on their surface (Allen and Nelson, 1910; Harvey, 1937 a).

The mechanism by which particles of ferric hydroxide are collected and adhere to the surface of plants is of interest.

Plants in suspension in fresh water are electronegative, colloidal ferric hydroxide in fresh water at pH 8 or more is electropositive. If ferric hydroxide sol in fresh water is protected by such colloids as gum arabic or albumin the charge is reversed (Harvey, 1937).

However, in sea water or salt solution the charges are dissipated (Hazel and Ayers, 1931).

These observations indicate that electrostatic attraction is unlikely. Whether micelles and particles just stick, on coming in contact with the plant's mucus-covered surfaces, or are adsorbed, perhaps on lipoid surfaces, is unknown. It may be pertinent that lecithin or olive oil collect ferric hydroxide from a protected sol.

The mechanism by which particulate iron enters the protoplasm of the plant cells is of interest, because the concentration of iron in true solution in sea water appears to be insufficient to supply the plants when growing rapidly.

In many marine unicellular species naked protoplasm is exposed to the water but in others there is an apparently complete exterior coat of cellulosic material, as in *Chlorella*. However, recent measurements of impedance across the exterior wall of a filament fresh-water alga by Bennett and Rideal (1954) indicates interstices in the cellulose covering. Hence it is possible that particles of ferric hydroxide are engulfed or ingested by the protoplasm; then they meet with acidic and reducing conditions which allow solution.

If, on the other hand, there are no interstices in the wall of *Chlorella*, it is possible to assume that slow solution takes place at the interfacial layer, which is a seat of strong electrochemical forces and behaves as if it had a greater hydrogen-ion concentration than the surrounding water. Slow solution of adherent particles may then supply sufficient diffusible ionic iron for the intracellular requirements of the organism. Stable organic compounds, as haematin or haemoglobin, are not utilized by the plants.

Evidence relating to the utilization of iron by marine plants and the effect of scarcity upon their growth is presented on p. 98.

Estimation. The determination of iron in sea water by means of thiocyanate is detailed in papers by Thompson and Bremner (1935), Seiwell (1935) and Braarud and Klem (1931). A method based on the red and violet association compounds formed by ferrous iron with $2:2'$-dipyridyl and $2:2':2''$-tripyridyl has been used by Cooper (1935). With the former reagent, quantities of iron in sea water down to less than 2 mg. Fe per m.³ could be ascertained. With the latter reagent less than 1 mg. Fe per m.³ could be detected. The salts present in sea water, other than iron, had no effect on the colour produced.

Besides determining iron soluble in strong hot acid, Cooper has determined the quantity of iron which dissolved in 24 hours in sea waters brought to pH 2·8 to which sulphite had been added. The relation which these quantities bear to the quantities available to plants is unknown. It might be possible to ascertain this by bioassay of a variety of sea waters, using *Nitzschia closterium*, which is readily grown until it becomes iron-deficient.

In order to interpret the results of analyses in terms of iron available to the plants, further investigation is needed. The parameters of interest are the number of particles or micelles the plants can collect and the readiness with which ionic iron is given off from such particles. If solution is on the surface of the plants where the pH is likely to be some 2 units lower than in the sea (Danielli, 1944), this second parameter may be the rate of solution at around pH 6. If the particles are ingested they pass into a locus which is both reducing and at a lower pH. In either event the rate of solution varies greatly with size and age of particle or micelle.

Manganese. In Pacific waters Thompson and Wilson (1935) found 1–10 mg. Mn per m.³ soluble in strong acids, and 0·07 % Mn in the ash of plankton, which was mostly plant organisms.

Pettersson (1945) concludes that there is a continuous rain of manganese oxides falling on the sea in the form of volcanic dust, and that in the deep oceans some 10 mg. fall annually on each metre of sea floor.

Although a manganous salt when added to sea water, even in quantity, is not precipitated, most of the manganese in the sea appears to be in the form of oxide particles with a very small and variable quantity present in solution. This opinion is bas d on estimates, in sea water buffered at 4·7, of the catalytic r ion of manganese in promoting the oxidation of tetramethyldiaminodiphenylmethane by periodate (Harvey, 1945).

In the English Channel, analyses showed 0·7–1·0 mg. of Mn soluble at pH 4·5 per m.³ After storage and sedimentation

of particulate matter in the water the concentration fell to
0·0–0·25 mg. Mn per m.^{3} A biological assay of manganese,
using *Chlamydomonas* (which can easily be obtained manganese-
deficient and responsive to the addition of 0·1 mg. Mn per m^3),
showed that this element in solution in sea water is readily
adsorbed on organic detritus (Harvey, 1947).   ⋆

It is not unlikely that the limited supply of available manganese may affect the growth of some phytoplankton species,
particularly in sea areas far from land. The following observation is suggestive. A marine *Chlamydomonas* which had been
in culture in artificial sea water, made from laboratory reagents,
over a period of many years, made a poor growth of noticeably
small cells when transferred to a sea water enriched with nitrogen,
phosphorus and iron. With the addition of soluble manganese,
heavy crops resulted.

Iodine. This element, occurring to the extent of some 50 mg.
I per m^3, is heavily concentrated by seaweeds. In *Laminaria* part
of the iodine occurs as di-iodo-tyrosine (Roche and Lafon, 1949). ⋆

Silicon. Silicon occurs in sea water as silicate, probably in
true rather than in colloidal solution. In the upper layers, where
it is utilized by diatoms, less than 10 mg. Si per m.^{3} have frequently been found. The deep water of the Atlantic contains
from about 1000 mg. Si per m^3 in the north (see fig. 61), while
in the Antarctic Clowes (1938) records over 3000 mg. Si per m.^{3}

When diatoms are eaten and the remains sink, silica dissolves
slowly into the water. In the Antarctic, where, during the short
summer, there is a heavy production of both thin-walled and
thick-walled (discoid) species, the latter are found as broken
frustules in the diatom ooze at a great depth. Presumably the
silica in the more numerous thin-walled species had dissolved
in transit. A culture of a thin-walled species, *Ditylum*, killed
by warming, and stored at pH *c.* 8·2 in a polythene bottle, returned 50 % of its frustule silica to the water in two months.

The seasonal changes in the silicate content of the water of
the English Channel have been followed by Atkins (1923–30),
Cooper (1933), and Armstrong (1951), who developed and
employed a precise method of determination. The concentra-

tions found throughout the water column during 1951 are shown in fig. 62. Rapid utilization of silicate occurred during the spring outburst of diatoms until the water was heavily depleted at the end of April. Then, starting without delay, re-solution became more rapid than utilization except in the upper 20 m. layer.

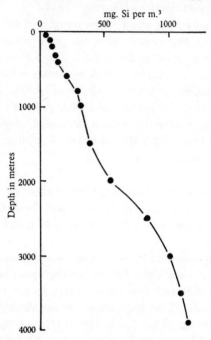

FIG. 61. Distribution of silicate with depth in the North Atlantic. 47° 24′ N., 7° 52′ W., May 1950. (From data by F. A. J. Armstrong, 1951.)

In addition to dissolved silica, the water, even far off land, contains a material quantity of silicon in particulate matter in suspension, in clay and presumably in undissolved diatom frustules.

The silicon in sea water which reacts on adding acid molybdate is assumed to be in the form of silicate. Cooper (1952) discusses the forms in which it may be present.

Chow and Robinson (1953) found that if a sol of colloidal silicic acid, which does not react with acid molybdate, was added to sea water, it slowly became reactive. Additions containing 25–40 mg. Si per litre were wholly converted within a week.

Since the concentration of silicate in ocean waters varies so greatly, and since it can be estimated rapidly and precisely (p. 198), its use is likely to increase in distinguishing water masses, in conjunction with temperature, salinity, and dissolved oxygen characteristics.

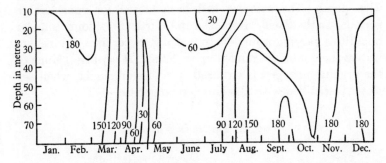

Fig. 62. Isopleth diagram showing the concentration of dissolved silicate (as mg. SiO_2 per m³) in the water column, which extends to a depth of 70 m., at a position in the English Channel 20 miles south-west of Plymouth from month to month during 1951 (Armstrong, 1954).

ORGANIC COMPOUNDS

There is a small quantity of dissolved organic matter in sea water, variable in amount and derived from the excreta of living organisms and from solution of their tissues when dead. A small amount may leach from living plants, but it is doubtful if this occurs to any material extent before the plants die (Krogh, Lange and Smith, 1930).

The large amount of sodium chloride present makes estimation difficult. Krogh and Keys (1934) have developed a method of direct combustion. A preliminary report of results obtained in this way (Keys, Christensen and Krogh, 1935) states that the organic matter in sea water is generally around 1·2–2 mg. C per litre with c. 0·2 mg. of organic nitrogen. In a later publication Krogh (1934) gives values obtained at a deep-water station in the Atlantic, 30° N., 69° W.   ★

Datzko (1951), using a more rapid method of carbon examination, found 2·9–3·2 mg. total C per litre in numerous analyses of waters from different depths in the Black Sea. This dissolved organic matter contains a small amount of phosphorus, and there are grounds for the inference that traces of organic sulphur compounds are also present.

The organic nitrogen in sea water has been estimated in the Atlantic by Von Brand and Rakestraw (1941), who found between 0·105 and 0·239 mg. N per litre, varying irregularly with depth. The samples had been stored for considerable periods. In the Pacific, Robinson and Wirth (1934) found smaller quantities varying from 0·03 to 0·12 mg. N per litre. In inshore waters the quantities ranged up to 0·3 mg. N per litre. Much of the organic nitrogen is oxidized, yielding ammonia, when sea water is distilled with alkaline permanganate.

Organic matter in sea water

Depth	Amount (mg. per litre)	
	Total carbon	Organic nitrogen
Surface	—	0·258
25 metres	2·40	—
1000	—	0·248
2000	2·35	0·240
3250	2·32	0·229
3750	2·36	—
4250	2·17	0·250
4750	2·48	0·238

The organic phosphorus in the water at a position in the North Atlantic has been found by Redfield, Smith and Ketchum (1937) to vary from nil to 0·016 mg. P per litre, the greatest quantities occurring in the upper layers during summer and autumn. The seasonal change in organic phosphorus in the water of the English Channel has been followed for several years at a position 20 miles offshore. Atlantic water from 1000 metres and greater depths contained no organic phosphorus (p. 44).

Variable quantities of an organic substance, having the characteristics of a reducing sugar, have been found in waters from the Gulf of Mexico. The greatest concentrations occurred in low salinity water. This substance affects the pumping rate of water by oysters (p. 120).

Yellow carotenoid substances, which fluoresce in ultra-violet light, have been extracted from coastal waters (Kalle, 1949; see also p. 84).

Evidence that sea waters contain vitamin B_{12} in concentrations having an order of magnitude around 0·01 mg. per m³ is given on p. 105. This vitamin is produced by many bacteria and is found concentrated in seaweeds.

The dissolved organic matter in sea water also contains surface-active compounds which absorb on solid surfaces, particularly at a meniscus (Harvey, 1941). This results in concentration of food for bacteria, which multiply rapidly when sea water is stored in small vessels or stored with sand which presents a large surface. When filtered sea water was stored in such vessels various observations indicated that 25–50 % of the dissolved organic matter was oxidized to carbon dioxide by bacteria (p. 71).

In the fresh water of Lake Mendota the total organic matter in solution ranges around 12 mg. per litre, about two to three times as much as in offshore sea water (Birge and Juday, 1934), and the total organic nitrogen ranges around 0·4 mg. per litre, ★ also some two to three times more than in offshore sea water. In the lake water between 60 and 80 % of the organic nitrogen was in the amino form and a quarter to a third precipitated by tannic and phosphotungstic acid. The water after freeing from organisms and evaporating gave positive reactions for proteins. Estimations after hydrolysis showed an average of 13 mg. of tryptophane, tyrosine and histidine and of 4 mg. per m³ of cystine (Domogalla, Juday and Peterson, 1925; Peterson, Fred and Domogalla, 1925).

Lake water has been found to contain biologically active compounds such as thiamine (0·15–0·9 mg. per m³) and biotin (0·0003–0·004 mg. per m³) by Hutchinson and Setlow (1946), and the cobalt-containing vitamin B_{12} (2 mg. per m³ in winter and 0·1 mg. per m³ in spring—Robbins, Hervey and Stebbins, 1950).

OXIDATION CATALYSTS AND INHIBITORS
IN SEA WATER

The fate of dissolved organic matter, its oxidation to carbon dioxide, is usually attributed to decomposition by bacteria, which reduce its concentration to the limit below which they cannot obtain sufficient food to make good their high basal metabolism. It seems reasonable to expect that a fraction undergoes oxidation, or partial oxidation, without the agency of bacteria. The magnitude of this fraction is quite unknown.

Some early experiments showed that in waters from different depths and positions, when brought to the same pH, various organic substances oxidized at very different rates. Added arsenite also oxidizes at different rates in different waters, and hydrogen peroxide decomposes at different rates (Harvey, 1925). In a series of papers Matsudaira (1950–2) has since presented a study of the catalytic action of sea waters towards hydrogen peroxide, and of negative catalysts which inhibit decomposition. He finds an inverse relation between rate of diatom growth in sea waters and their catalytic activity towards hydrogen peroxide.

It is only when the dissolved oxygen concentration in the water falls to extremely low values that the oxidation-reduction potential of sea water falls below about $+0\cdot3$ V. (Cooper, 1937); sea water contains traces of many substances which are likely to 'trigger' or catalyse oxidations.

Evidence suggesting that enzymes occur in solution in sea water is discussed on p. 67.

CHAPTER X

THE CARBON DIOXIDE SYSTEM

SEA WATER contains carbon dioxide as bicarbonate and carbonate ions, as undissociated molecules of CO_2 and as carbonic acid, all in equilibrium with each other, and with the hydrogen ions present:

$$\begin{array}{c} CO_2 \\ \Updownarrow \\ H_2CO_3 \end{array} \rightleftharpoons HCO_3' + H^{\cdot} \rightleftharpoons CO_3'' + H^{\cdot}.$$

The water contains cations in excess of the equivalent anions derived from strong acids. This excess base, or titration alkalinity, is equivalent to that of the bicarbonate carbonate and borate ions in the water. Weak acids, other than carbonic and boric, are present in such small quantities in water from the open sea that they do not come into the picture.

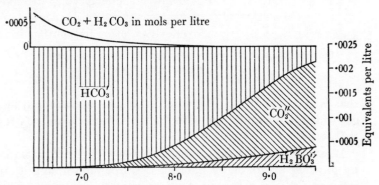

FIG. 63. Change in constitution of an ocean water with changing pH. Temperature 16° C., salinity 36 ‰, excess base 0·00246 equivalent per litre.

The diagram (fig. 63) shows the change in these constituents with varying hydrogen-ion concentration for a typical ocean water of salinity 36‰, at 16° C., containing 0·00246 equivalent of excess base per litre.

The free (undissociated) carbon dioxide consists of CO_2 and H_2CO_3 molecules in solution and in equilibrium with each other.

The quantity of H_2CO_3 is about 1 % of the quantity of CO_2 molecules. This free carbon dioxide exerts a partial pressure. When sea water is shaken with a bubble of inert gas for 5–10 minutes, carbon dioxide passes into the bubble, which will have attained a partial pressure of carbon dioxide equal to that exerted by the water.

Fig. 64 shows how the partial pressure of carbon dioxide exerted by an ocean water changes with the pH of the water,

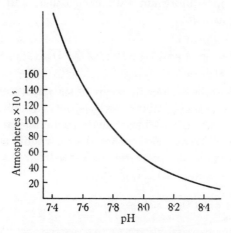

FIG. 64. The partial pressure of carbon dioxide in atmospheres $\times 10^5$
in an ocean water of 36 ‰ S at 16° C.

that is, how the partial pressure of carbon dioxide in an inert gas which is in equilibrium with the water varies with the pH of the water.

If carbon dioxide is abstracted from sea water as, for instance, by growing plants or by blowing CO_2-free air through it, the following change takes place: bicarbonate ions change to carbonate; the pH increases and in consequence boric acid in solution dissociates; the molecular CO_2 in solution diminishes and its partial pressure declines. A particular example, given on p. 5, shows the quantities involved.

It is possible to calculate the total carbon dioxide content, its partial pressure and the concentration of bicarbonate or carbonate ions for a sea water of known salinity, temperature, pH and excess base. If one of these variables is altered a new state of equilibrium within the water system is attained within a few minutes. Transference of carbon dioxide to or from the air, however, takes place relatively slowly.

HYDROGEN-ION CONCENTRATION

The hydrogen-ion concentration in the upper layers of the open oceans varies within rather narrow limits. Values as low as pH 8 are rarely met with and the upper limit rarely exceeds pH 8·3. In shallow waters and rock pools the range is greater; plants utilize carbon dioxide and raise the pH, while the respiration of organisms acts in the opposite direction. At great depths in the oceans the pH is lowered owing to the effect of pressure; in the South Atlantic Wattenberg (1933) records values between pH 8·1 and 7·9 *in situ*. Intermediate layers are found in some areas where the decomposition of organisms falling from above is taking place; off the west coast of Africa values of pH 7·6 *in situ* are recorded in such a layer at 400 m. depth.

TABLE 18. *Change in* pH *of sea water for rise of* 1° C. (*x*)

(From Buch and Nynäs, 1939.)

pH	Cl ‰ = 10			Cl ‰ = 15		
	0–20°	10–20°	20–30°	0–20°	10–20°	20–30°
7·4	− 0·0087	− 0·0084	− 0·0069	− 0·0088	− 0·0087	− 0·0076
7·6	92	92	79	95	96	83
7·8	100	101	89	103	105	90
8·0	108	109	94	110	112	94
8·2	114	115	98	115	117	96
8·4	117	117	99	118	118	98

pH	Cl ‰ = 19·5			Cl ‰ = 21		
	0–20°	10–20°	20–30°	0–20°	10–20°	20–30°
7·4	− 0·0089	− 0·0087	− 0·0081	− 0·0092	− 0·0089	− 0·0079
7·6	95	95	91	97	98	88
7·8	104	104	98	106	108	93
8·0	110	109	102	112	114	96
8·2	114	112	103	116	116	98
8·4	116	114	104	118	119	100

Note. The table contains some irregular values.

The hydrogen-ion concentration of any sample of sea water depends primarily upon the concentration of carbon dioxide and of the neutral salts. It is influenced by both the temperature of the water and by great pressure.

The effect of temperature on the hydrogen-ion concentration. The hydrogen-ion concentration increases (pH falls) with rising temperature, the increase depending upon the pH and salinity of the water. Table 18 shows this relationship under conditions where there is no interchange of carbon dioxide with the atmosphere. The values, copied from the original publication, show some minor irregularities. The values (x) in this table allow solution of the following equation:

$$pH_{t_2} = pH_{t_1} + x(t_2 - t_1),$$

where pH_{t_2} is the pH of the water at t_2 and pH_{t_1} is the pH at t_1.

The effect of pressure on the hydrogen-ion concentration. When sea water is subjected to pressure at great depths, the dissociation constants of carbonic acid are altered and as a result the pH of the water is decreased (p. 180). From investigations of the change in these constants, the change in pH has been calculated (Buch and Gripenberg, 1932):

pH at atmo-spheric pressure	ΔpH for increase of 1000 m. in depth
7·5	− 0·035
7·6	− 0·031
7·7	− 0·028
7·8	− 0·025
7·9	− 0·023
8·0	− 0·022
8·1	− 0·021
8·2	− 0·020
8·3	− 0·020

Measurement of hydrogen-ion concentration. The hydrogen-ion concentration of sea water has been measured directly by means of the quinhydrone electrode for pH values up to about pH 8; the use of a hydrogen electrode is restricted, since the stream of gas washes carbon dioxide out of the water, causing a gradual rise in pH. Above pH 8, measurements are made by comparison with buffer solutions whose pH has been determined by means of the hydrogen electrode and is known for the temperature at which the comparison is made. This comparison is effected either by means of indicators or by means of the glass electrode.

The hydrogen-ion concentration of sea water is commonly determined by means of indicators against buffer solutions, the colour comparison being made visually or with a spectrophotometer or a photoelectric colorimeter; either instrument is stated to give values within $0.01-0.02$ pH. It is necessary to take into consideration the salt error of the indicator, and the effect of the temperature at which colour comparison is made both upon the buffer solution and upon the indicator.

Salt error of indicators

Salinity (‰)	5	10	15	20	25	30	35
Cresol red ⎫ Xylenol blue ⎭	-0.04	-0.12	-0.17	-0.21	-0.23	-0.25	-0.26
Thymol blue	-0.04	-0.11	-0.16	-0.19	-0.21	-0.22	-0.23
Phenol red	-0.06	-0.15	-0.21	-0.24	-0.26	-0.28	-0.29

(From Buch and Nynäs, 1939.)

The salt error has been investigated for the commonly used indicators, against Palitzsch borate-boric acid buffers. It is a minus value; the corrected pH of the sea water is less than that of the buffer having the same colour. McClendon (1917) buffer solutions calibrated at $20°$ C. are designed to have the same salt effect as ocean waters upon the indicator and hence dispense with this correction. However, they have the disadvantage that salts crystallize out from the solutions at low temperatures.

The effect of temperature on Palitzsch borax-borate buffer mixtures has been calculated by Buch (1929). The temperature coefficients (y), being the change in pH of the buffer solution for a rise of $1°$ C., are shown in the table on p. 158.

It is usual to make the colour comparison when the sample of sea water and the buffer solutions are at the same temperature (t_1). Under these circumstances, where Palitzsch buffers are used, the corrected pH of the sea water at the temperature of observation $=$ pH observed $+ y(t_1 - 18°) +$ 'salt error', and

$$pH_{t_w} = pH \text{ observed} + x(t_w - t_1) + y(t_1 - 18) + \text{'salt error'},$$

158   THE CARBON DIOXIDE SYSTEM

where pH_{t_w} is the pH of the sea water *in situ* at $t_w°$ C. It is to be noted that x, y and 'salt error' are each minus quantities.

If the temperature of the water sample differs from that of the buffer at the time of comparison, the effect of temperature on the dissociation of the indicator has also to be taken into consideration.

Palitzsch borax-borate buffer mixtures

Borax solution: 19·108 g. $Na_2B_4O_7$, $10H_2O$ per litre.
Boric acid solution: 12·404 g. boric acid + 2·925 g. NaCl per litre.

Borax solution (cm³)	Boric solution (cm³)	pH at 18°C.	Temperature coefficient (y)
6·0	4·0	8·69	−0·0058
5·5	4·5	8·60	−0·0054
5·0	5·0	8·51	−0·0050
4·5	5·5	8·41	−0·0048
4·0	6·0	8·31	−0·0045
3·5	6·5	8·20	−0·0042
3·0	7·0	8·08	−0·0039
2·5	7·5	7·94	−0·0037
2·3	7·7	7·88	−0·0035
2·0	8·0	7·78	−0·0033
1·5	8·5	7·60	−0·0030
1·0	9·0	7·36	−0·0026
0·6	9·4	7·09	−0·0022

The glass electrode is coming into increasing use for the measurement of hydrogen-ion concentration in sea water. The method is only comparative, because the 'asymmetry potential' of the electrode varies and this necessitates standardizing the apparatus against a buffer solution before and after making a series of determinations. A study of the glass electrode method of comparison against the indicator method has been made by Buch and Nynäs (1939), who considered the accuracy similar to that obtained by indicators using a photoelectric colorimeter for comparison—a method to which they were inclined to give preference and whose accuracy they estimate at 0·01–0·02 pH unit. There is no 'salt error' in estimations made with the glass electrode; the temperature coefficient of the buffer solution used for standardizing has to be taken into account; the temperature coefficient of sea water (Table 2) allows the pH *in situ* to be calculated from the pH at temperature of observation.

THE DISSOCIATION CONSTANT (IONIC PRODUCT) OF WATER AND THE HYDROXYL-ION CONCENTRATION

The hydroxyl-ion concentration, C_{OH_1} in gram-mols per litre equals K_w/C_H, where K_w is the (molal) dissociation constant or ionic product of water. This value changes with temperature and with the salt content of sea water.

Effect of temperature on the ionic product of pure water

$T°$ C.	K_w or $C_H \times C_{OH}$		pK_w or $pH + pOH$	
0	$0·12 \times 10^{-14}$	14·92		Kohlrausch and Heydweiller
	0·14	14·85		Lorentz and Böhi
16	0·63	14·20		Michaelis
18	0·74	14·130		Michaelis
20	0·86	14·065		Michaelis
22	1·01	13·995		Michaelis
24	1·19	13·925		Michaelis

The ionic product of water in sodium chloride solutions has been investigated by Bjerrum and Unmack (1929), and, from the relations found by them, Buch (1938) has interpolated values for sea water. The result indicates that the molal ionic product in ocean water is about 1·75 times the value in fresh water.

From these values, the neutral point of ocean water, where the concentration of hydroxyl ions equals that of hydrogen ions, lies at pH 7·33 at 0° C., at pH 6·98 at 16° C. and at pH 6·84 at 24° C.

Bjerrum and Unmack (1929) have also investigated the activity of hydroxyl ions in salt solutions. These results indicate that the activity coefficient f_{OH} in ocean water is c. 0·945.

Cooper (1937) has made computations for sea water using the thermodynamic dissociation product

$$K_w = \frac{a_{H·} \times a_{OH'}}{a_{H_2O}},$$

where a stands for the activity of the respective ion or molecule and $a_x = f_x c_x$, where f is the activity coefficient of x, and c_x its molar concentration.

The thermodynamic product for pure water, where a_{H_2O} is unity, has been determined by Harned and co-workers. In sea water the effect of salts is to change the activity of the undissociated water, and of the hydroxyl ions. The pH as measured electrometrically, either direct or via indicators, is in fact, a measurement of the activity and is more correctly pa_H. The effect of salts in sea water on the activity of the undissociated water, a_{H_2O}, has been found from the lowering of its freezing-point or its vapour pressure (p. 169).

Hence the (thermodynamic) ionic product $(a_H \times a_{OH})$ for sea water $= K_w a_{H_2O}$, where a_{H_2O} is the activity of the water molecules in the sea water.

$t°$ C.	Water $pa_H + pa_{OH}$	Sea water $(Cl = 19‰)$ $pa_H + pa_{OH}$
0	14·939	14·947
5	14·731	14·739
10	14·533	14·541
15	14·345	14·353
20	14·167	14·175
25	13·997	14·005
30	13·832	13·840

TITRATION ALKALINITY, SPECIFIC ALKALINITY AND CARBONATE ALKALINITY

The excess base, 'alkali reserve' or *titration alkalinity* is found by titrating sea water with a strong acid. Since the normality of a natural sea water rarely exceeds $0·0026N$ ($260 \times 10^{-5}N$), it is desirable to adhere to a particular technique in order to obtain strictly comparable values. The following method (Wattenberg, 1933, p. 139) is in general use. A stream of air, washed free from CO_2, is passed through 200 cm³ of the sample of sea water to which 10 cm³ of $N/20$ HCl has been added and the liquid heated to boiling, in order to drive off all the CO_2 set free. The hot liquid is then back titrated with $N/20$ barium hydroxide using a mixture of 3 parts of Brom cresol green and 2 parts of methyl red as indicator, which gives a sharp end-point at pH 5. The
★ value so obtained is expressed in equivalents per litre.

Gripenberg (1937) points out that titration to pH 5 rather than to neutrality gives values which are too high by a small systematic error, and recommends titration to an end-point between pH 6 and 7. Moreover, the precipitate formed during titration with barium hydroxide has a slight effect on the indicator, for which reason titration with carbonate-free sodium hydroxide is advocated.

There is, in general, a linear relation between titration alkalinity and the total salt content of the water. With some exceptions, the following relation is a close approximation:

$$\frac{\text{Titration alkalinity} \times 10^3}{\text{Cl}\%_0} = 0.123 \ \text{(the } \textit{specific alkalinity}).$$

At great depths, calcium carbonate is dissolved from the bottom deposits and the specific alkalinity of the water layer close to the bottom is greater. In some areas calcareou. organisms absorb calcium from the water and the specific alkalinity of the upper layers is reduced.

Wattenberg (1933) has determined the specific alkalinity of the water at a number of positions in the Atlantic. Mean values at various depths are shown in the following table:

Depth (m.)	Specific alkalinity
0	0.1209
50	0.1207
100	0.1204
200	0.1210
250	0.1210
1000	0.1225
2500	0.1225
Close to bottom at c. 2000	0.1225
3000	0.1241
4000	0.1250
5000	0.1267

A convenient method of estimating alkalinity, due to Thompson and Anderson (1940) has been used by Bruneau, Jerlov and Koczy (1953), who confirmed earlier observations that specific alkalinity tended to be greater in the Pacific than in the Atlantic.

It is convenient to express titration alkalinity in terms of 10^{-5} normality, then

$$\text{Titration alkalinity} = 12 \cdot 3 \times \text{Cl}\%_0 \times 10^{-5}.$$

This relation only holds for waters which fall into neither of the above categories. Furthermore, in low salinity Baltic waters Gripenberg (1937) has found marked variations in the ratio between excess base and salinity.

Since the only weak acids in sea water which are known to be present in sufficient quantity to affect the equilibria are carbonic and boric

$$\text{Titration alkalinity} = C_{\text{HCO}_3'} + 2C_{\text{CO}_3''} + C_{\text{H}_2\text{BO}_3'} + (C_{\text{OH}'} - C_{\text{H}\cdot}),$$

where $C_{\text{HCO}_3'}$, etc., is the concentration of the ion in gram-ions per litre.

The term *carbonate alkalinity* is used for that part of the titration alkalinity which is equivalent to the bicarbonate and carbonate ions:

$$\begin{aligned} \text{Carbonate alkalinity} &= C_{\text{HCO}_3'} + 2C_{\text{CO}_3''} \\ &= \text{Titration alkalinity} - C_{\text{H}_2\text{BO}_3'} \\ &\qquad\qquad - (C_{\text{OH}'} - C_{\text{H}\cdot}). \end{aligned}$$

It remains to find the value of $C_{\text{H}_2\text{BO}_3'}$.

For boric acid in pure water at infinite dilution

$$\frac{C_{\text{H}\cdot} \times C_{\text{H}_2\text{BO}_3'}}{C_{\text{H}_3\text{BO}_3}} = K_{\text{B}},$$

the dissociation constant for boric acid. This constant increases with temperature. In sea water the equilibrium is affected by the salts present. A table has been drawn up by Buch (1933), giving values of the 'apparent' dissociation constant, K_{B}', for various temperatures and chloride contents. The apparent constant K_{B}' for boric acid is given in table 19.

The boron content of sea water is found to bear a direct linear relation to the chloride present. The total boron $C_{\Sigma \text{B}}$, or $C_{\text{H}_2\text{BO}_3'} + C_{\text{H}_3\text{BO}_3}$, approximates to $2 \cdot 2 \times 10^{-5} \times \text{Cl}\%_0$ mols per litre. Hence

$$\begin{aligned} \text{Carbonate alkalinity} &= \text{titration alkalinity} \\ &\quad - \frac{K_{\text{B}}' \times C_{\Sigma \text{B}}}{C_{\text{H}\cdot} + K_{\text{B}}'} - (C_{\text{OH}'} - C_{\text{H}\cdot}) \end{aligned}$$

TABLE 19. *Apparent dissociation constant* K'_B *of boric acid in sea water*

Cl ‰	°C / S‰	0	2	4	6	8	10	12	14	16	18	20	22	24	26	28	30
15	27.1	1.10	1.15	1.20	1.23	1.29	1.35	1.41	1.44	1.51	1.55	1.62	1.70	1.78	1.82	1.91	1.95×10^{-9}
16	28.9	1.12	1.17	1.23	1.29	1.35	1.38	1.44	1.51	1.55	1.62	1.70	1.74	1.82	1.91	1.95	2.00
17	30.7	1.17	1.23	1.29	1.32	1.38	1.44	1.51	1.55	1.62	1.70	1.74	1.82	1.91	1.95	2.04	2.09
18	32.5	1.20	1.26	1.32	1.38	1.44	1.48	1.55	1.62	1.66	1.74	1.82	1.86	1.95	2.00	2.09	2.14
19	34.3	1.26	1.32	1.35	1.41	1.48	1.55	1.62	1.66	1.74	1.82	1.86	1.91	2.00	2.09	2.14	2.24
20	36.1	1.29	1.35	1.41	1.48	1.55	1.58	1.66	1.74	1.78	1.86	1.95	2.00	2.09	2.14	2.24	2.29
21	37.9	1.32	1.38	1.44	1.51	1.58	1.66	1.70	1.82	1.86	1.91	2.00	2.09	2.14	2.24	2.29	2.40
0	$K_B =$	0.40	0.42	0.44	0.46	0.48	0.50	0.52	0.54	0.56	0.58	0.60	0.63	0.65	0.67	0.69	0.72×10^{-9}

TABLE 20. *Quantities to be subtracted from the titration alkalinity in order to obtain the carbonate alkalinity*

(From Buch, 1951.)

Cl = 15 ‰

°C. pH	0	2	4	6	8	10	12	14	16	18	20	22	24	26	28	30
7·4	1	1	1	1	1	1	1	1	1	1	1	1	1	1	1	2×10^{-5}
7·5	1	1	1	1	1	1	1	1	2	2	2	2	2	2	2	2
7·6	1	1	1	2	2	2	2	2	2	2	2	2	2	2	2	2
7·7	2	2	2	2	2	2	2	2	2	2	2	3	3	3	3	3
7·8	2	2	2	2	2	3	3	3	3	3	3	3	3	3	3	4
7·9	3	3	3	3	3	3	3	3	4	4	4	4	4	4	4	4
8·0	3	3	4	4	4	4	4	4	4	5	5	5	5	5	5	5
8·1	4	4	4	4	5	5	5	5	5	5	5	6	6	6	5	6
8·2	5	5	5	5	6	6	6	6	6	6	6	6	6	7	6	8
8·3	6	6	6	7	7	7	7	7	8	8	7	8	7	9	8	9
8·4	7	7	8	8	8	8	9	9	9	9	8	10	9	10	11	11
8·5	8	9	9	9	10	10	10	10	11	11	11	11	12	12	12	13×10^{-5}

Cl = 17 ‰

°C. pH	0	2	4	6	8	10	12	14	16	18	20	22	24	26	28	30
7·4	1	1	1	1	1	1	1	1	1	2	2	2	2	2	2	2×10^{-5}
7·5	1	1	1	2	2	2	2	2	2	2	2	2	2	2	2	2
7·6	2	2	2	2	2	2	2	2	2	3	3	3	3	3	3	3
7·7	2	3	3	3	3	3	3	3	3	3	3	3	3	3	3	4
7·8	3	3	3	3	3	4	4	4	4	4	4	4	4	4	4	4
7·9	4	4	4	4	5	5	5	5	5	5	5	5	5	5	5	5
8·0	5	5	5	5	6	6	6	6	6	7	6	6	6	6	6	6
8·1	6	6	6	6	7	7	7	7	6	7	7	7	7	8	8	8
8·2	7	7	8	8	8	8	9	9	9	9	8	8	9	9	9	9
8·3	8	9	9	8	10	10	10	11	11	11	10	10	10	11	11	11
8·4	10	10	11	9	11	12	12	12	13	11	11	12	12	12	13	13
8·5	10	10	11	11	12	12	12	13	13	13	13	14	14	14	15	15×10^{-5}

TABLE 20 (cont.)

Cl = 19 ‰

°C. pH	0	2	4	6	8	10	12	14	16	18	20	22	24	26	28	30
7·4	1	1	1	1	2	2	2	2	2	2	2	2	2	2	2	2×10^{-5}
7·5	2	2	2	2	2	2	2	2	2	2	2	2	3	3	3	3
7·6	2	2	2	2	2	2	3	3	3	3	3	3	3	3	3	3
7·7	2	3	3	3	3	3	3	3	3	3	4	4	4	4	4	4
7·8	3	3	3	3	4	4	4	4	4	4	4	5	5	5	5	5
7·9	4	4	4	4	4	5	5	5	5	5	5	6	6	6	6	6
8·0	5	5	5	5	5	6	6	6	6	6	7	7	7	7	7	8
8·1	6	6	6	6	7	7	7	7	8	8	8	8	8	9	9	9
8·2	7	7	7	8	8	8	9	9	9	9	10	10	10	10	11	11
8·3	8	9	9	9	10	10	10	10	11	11	11	12	12	12	13	13
8·4	10	10	11	11	11	12	12	12	13	13	13	14	14	14	15	15
8·5	12	12	13	13	13	14	14	14	15	15	16	16	16	17	17	17×10^{-5}

Cl = 21 ‰

°C. pH	0	2	4	6	8	10	12	14	16	18	20	22	24	26	28	30
7·4	1	2	2	2	2	2	2	2	2	2	2	2	2	2	3	3×10^{-5}
7·5	2	2	2	2	2	2	2	2	3	3	3	3	3	3	3	3
7·6	2	2	2	3	3	3	3	3	3	3	3	4	4	4	4	4
7·7	3	3	3	3	3	4	4	4	4	4	4	4	4	5	5	5
7·8	4	4	4	4	4	4	5	5	5	5	5	5	6	6	6	6
7·9	4	4	5	5	5	5	6	6	6	6	6	7	7	7	7	7
8·0	5	6	6	6	6	7	7	7	7	7	8	8	8	8	9	9
8·1	7	7	7	7	9	8	8	8	9	9	9	10	10	10	10	11
8·2	8	8	9	9	9	10	10	10	10	11	11	11	12	12	12	13
8·3	10	10	10	11	11	11	12	12	12	13	13	14	14	14	15	15
8·4	11	11	11	12	12	13	13	13	14	14	14	15	15	15	16	16
8·5	14	14	14	15	15	16	16	17	17	17	18	18	19	19	20	20×10^{-5}

or

Carbonate alkalinity = titration alkalinity

$$-\frac{K'_B \times 2\cdot2 \times Cl\,^o/_{oo} \times 10^{-5}}{C_{H\cdot} + K'_B}$$
$$-(C_{OH'} - C_{H\cdot}) \text{ equivalents per litre}$$
$$\text{(Equation 1)}$$

Within the range of pH 5·5–8·5, the value of $C_{OH'} - C_{H\cdot}$ is negligible.

Values for the term $\dfrac{K'_B \times 2\cdot2 \times Cl\% \times 10^{-5}}{C_{H\cdot} + K'_B}$ for sea waters, with the range of pH 7·4–8·5 and 27–38‰ salinity and 0–30° C. are shown in table 20. By subtracting the appropriate value from the titration alkalinity, the carbonate alkalinity is obtained.

RELATIONS BETWEEN BICARBONATE, CARBONATE, CARBONIC ACID AND HYDROGEN-ION CONCENTRATION

In pure water at great dilution, the first dissociation

$$H_2CO_3 \rightleftharpoons H\cdot + HCO'_3$$

takes up an equilibrium where

$$\frac{C_{H\cdot} \times C_{HCO_3'}}{C_{H_2CO_3}} = K_1 \quad \text{(the \textit{first dissociation constant})}$$

for the particular temperature under consideration. The second dissociation, $HCO'_3 \rightleftharpoons H\cdot + CO''_3$, takes up an equilibrium where

$$\frac{C_{H\cdot} \times C_{CO_3''}}{C_{HCO_3'}} = K_2 \quad \text{(the \textit{second dissociation constant}).}$$

In sea water the concentrations of these ions are material. Furthermore, the neutral salts in the water reduce the *activity* of the water and permit the formation of complexes. Then

$$K_1 = \frac{a_{H\cdot} \times a_{HCO_3'}}{a_{H_2CO_3}} \quad \text{and} \quad K_2 = \frac{a_{H\cdot} \times a_{CO_3''}}{a_{HCO_3'}},$$

where a stands for the activity of the respective ion or molecule.

TABLE 21. *The first apparent dissociation constant of carbon dioxide in sea water,* K_1'

(From Buch, 1951.)

°C / Cl‰	0	2	4	6	8	10	12	14	16	18	20	22	24	26	28	30
15	0·58	0·62	0·65	0·68	0·71	0·74	0·77	0·80	0·83	0·86	0·89	0·92	0·95	0·97	0·99	1·01 × 10⁻⁶
16	0·59	0·63	0·66	0·69	0·72	0·75	0·78	0·81	0·84	0·87	0·91	0·93	0·96	0·99	1·01	1·03
17	0·60	0·64	0·67	0·70	0·73	0·76	0·79	0·82	0·86	0·89	0·92	0·95	0·98	1·00	1·02	1·05
18	0·61	0·65	0·68	0·71	0·74	0·77	0·81	0·84	0·87	0·90	0·93	0·97	0·99	1·02	1·04	1·06
19	0·62	0·66	0·69	0·72	0·76	0·79	0·82	0·85	0·88	0·92	0·95	0·98	1·01	1·04	1·06	1·08
20	0·63	0·67	0·70	0·73	0·77	0·80	0·83	0·87	0·90	0·93	0·97	1·00	1·03	1·05	1·07	1·10
21	0·64	0·68	0·71	0·75	0·78	0·81	0·84	0·88	0·91	0·95	0·98	1·01	1·04	1·07	1·09	1·12
0	0·26	0·28	0·29	0·31	0·32	0·34	0·35	0·37	0·38	0·39	0·40	0·42	0·43	0·44	0·44	0·45 × 10⁻⁶

TABLE 22. *The second apparent dissociation constant of carbon dioxide in sea water,* K_2'

(From Buch, 1951.)

°C / Cl‰	0	2	4	6	8	10	12	14	16	18	20	22	24	26	28	30
15	0·43	0·46	0·49	0·53	0·56	0·60	0·63	0·66	0·69	0·73	0·76	0·79	0·83	0·86	0·90	0·93 × 10⁻⁹
16	0·46	0·49	0·53	0·56	0·60	0·63	0·67	0·71	0·74	0·78	0·81	0·85	0·88	0·92	0·96	0·99
17	0·48	0·51	0·55	0·59	0·63	0·67	0·71	0·74	0·78	0·82	0·86	0·90	0·93	0·97	1·01	1·05
18	0·51	0·55	0·59	0·63	0·67	0·71	0·75	0·79	0·83	0·87	0·91	0·95	0·99	1·03	1·07	1·11
19	0·54	0·58	0·62	0·66	0·71	0·75	0·79	0·83	0·88	0·92	0·96	1·01	1·05	1·10	1·14	1·18
20	0·57	0·62	0·66	0·71	0·75	0·80	0·84	0·89	0·93	0·98	1·02	1·07	1·12	1·16	1·21	1·26
21	0·60	0·65	0·69	0·74	0·79	0·84	0·89	0·93	0·98	1·03	1·08	1·13	1·18	1·23	1·28	1·33 × 10⁻⁹
0	0·23	0·25	0·27	0·29	0·30	0·32	0·34	0·36	0·38	0·40	0·42	0·44	0·46	0·48	0·50	0·51 × 10⁻¹⁰

TABLE 23. *Solubility of carbon dioxide in pure water, α_0 and in sea waters, α_s, in mols per litre*

(From Bohr's data (Buch, 1951).)

°C / Cl‰	0	2	4	6	8	10	12	14	16	18	20	22	24	26	28	30
0	$\alpha_0=$ 770	712	662	619	576	536	502	472	442	417	394	372	351	332	314	299×10^{-4}
15	$\alpha_s=$ 674	623	578	538	504	472	442	416	393	371	351	331	314	299	284	270
16	667	617	573	533	499	468	438	413	390	368	348	329	312	297	281	268
17	660	611	567	528	495	464	434	410	387	365	346	327	310	294	279	266
18	653	605	562	524	490	460	431	406	384	362	343	324	307	292	277	264
19	646	599	557	519	486	456	428	403	381	359	340	321	304	289	275	262
20	640	593	551	514	482	452	424	400	377	356	337	319	302	287	273	260
21	633	587	546	509	477	448	421	396	374	354	335	317	300	285	271	258

For working purposes, 'apparent' dissociation constants, K_1' and K_2', have been used, where

$$K_1' = \frac{a_{\text{H}^{\cdot}} \times C_{\text{HCO}_3'}}{a_{\text{H}_2\text{CO}_3}} \quad \text{and} \quad K_2' = \frac{a_{\text{H}^{\cdot}} \times C_{\text{CO}_3''}}{C_{\text{HCO}_3'}}.$$

These two equations define the constants which are given in tables 21 and 22.

The unbound CO_2 in solution consists of free CO_2 with a small proportion of undissociated H_2CO_3. The term $a_{\text{H}_2\text{CO}_3}$ (the activity of the undissociated carbonic acid) has been taken as equal to $a_{\text{CO}_2} \times a_{\text{H}_2\text{O}}$.

From the definition of the activity of a volatile dissolved substance

$$a_{\text{CO}_2} = P_{\text{CO}_2} \times \alpha_0,$$

where P_{CO_2} is the partial pressure of CO_2 in atmospheres and α_0 its solubility in (pure) water in mols per litre at 1 atmosphere pressure at the particular temperature under consideration (table 23).

$a_{\text{H}_2\text{O}}$ is unity for pure water and is depressed by the presence of salts in solution. For any particular sea water it equals the ratio of its vapour pressure to that of pure water. This ratio has been determined and found to follow the equation

$$\frac{\text{V.P.}_{\text{sea water}}}{\text{V.P.}_{\text{water}}} = 1 - 0 \cdot 000969 \text{ Cl}\%_0.$$

The value of $a_{\text{H}_2\text{O}}$ may also be derived from the depression of the freezing-point of sea water below that of pure water, Δt°, by means of the Lewis and Randall equation

$$\log_{10} a_{\text{H}_2\text{O}} = -0 \cdot 004211 \Delta t - 0 \cdot 0000022 \Delta t^2.$$

TABLE 24. *The activity of water in sea water of varying salinity*

$$a_{\text{H}_2\text{O}} = \frac{\text{V.P.}_{\text{sea water}}}{\text{V.P.}_{\text{fresh water}}}$$

Cl$\%_0$	$a_{\text{H}_2\text{O}}$	Cl$\%_0$	$a_{\text{H}_2\text{O}}$
2	0·998	14	0·986
4	0·996	16	0·984
6	0·994	18	0·983
8	0·992	20	0·981
10	0·990	22	0·979

TABLE 25. *Factors which when multiplied by the carbonate alkalinity yield the value of the partial pressure of carbon dioxide in atmospheres*

(From Buch 1951.)

Cl = 15‰

°C / pH	0	2	4	6	8	10	12	14	16	18	20	22	24	26	28	30
7·4	0·88	0·90	0·92	0·94	0·97	0·99	1·01	1·04	1·06	1·08	1·11	1·13	1·16	1·20	1·24	1·28
7·5	0·70	0·71	0·73	0·75	0·76	0·78	0·80	0·82	0·84	0·85	0·87	0·89	0·91	0·94	0·96	1·00
7·6	0·55	0·56	0·57	0·59	0·60	0·61	0·63	0·64	0·66	0·67	0·68	0·70	0·72	0·74	0·76	0·79
7·7	0·43	0·44	0·45	0·46	0·47	0·48	0·49	0·50	0·52	0·53	0·54	0·55	0·56	0·58	0·60	0·61
7·8	0·34	0·35	0·35	0·36	0·37	0·38	0·38	0·39	0·40	0·41	0·42	0·43	0·44	0·45	0·46	0·48
7·9	0·27	0·27	0·28	0·28	0·29	0·29	0·30	0·31	0·31	0·32	0·33	0·33	0·34	0·35	0·36	0·37
8·0	0·21	0·21	0·22	0·22	0·22	0·23	0·23	0·24	0·24	0·25	0·25	0·26	0·26	0·27	0·28	0·28
8·1	0·16	0·16	0·17	0·17	0·17	0·18	0·18	0·18	0·19	0·19	0·19	0·20	0·20	0·20	0·21	0·22
8·2	0·12	0·13	0·13	0·13	0·13	0·14	0·14	0·14	0·14	0·14	0·15	0·15	0·15	0·16	0·16	0·17
8·3	0·09	0·10	0·10	0·10	0·10	0·10	0·10	0·11	0·11	0·11	0·11	0·11	0·11	0·12	0·12	0·12
8·4	0·07	0·07	0·08	0·08	0·08	0·08	0·08	0·08	0·08	0·08	0·08	0·09	0·09	0·09	0·09	0·09
8·5	0·05	0·05	0·06	0·06	0·06	0·06	0·06	0·06	0·06	0·06	0·06	0·06	0·06	0·06	0·07	0·07

Cl = 17‰

°C / pH	0	2	4	6	8	10	12	14	16	18	20	22	24	26	28	30
7·4	0·85	0·87	0·89	0·92	0·94	0·96	0·98	1·00	1·02	1·05	1·07	1·09	1·12	1·16	1·19	1·23
7·5	0·67	0·69	0·70	0·72	0·74	0·75	0·77	0·79	0·81	0·83	0·84	0·86	0·88	0·91	0·94	0·97
7·6	0·53	0·54	0·56	0·57	0·58	0·60	0·61	0·62	0·63	0·65	0·66	0·67	0·69	0·71	0·74	0·76
7·7	0·41	0·42	0·43	0·45	0·46	0·47	0·48	0·49	0·50	0·51	0·52	0·53	0·54	0·56	0·57	0·59
7·8	0·33	0·33	0·34	0·35	0·35	0·36	0·37	0·38	0·38	0·39	0·40	0·41	0·42	0·43	0·44	0·45
7·9	0·25	0·26	0·27	0·27	0·28	0·28	0·29	0·29	0·30	0·31	0·31	0·33	0·33	0·33	0·34	0·35
8·0	0·20	0·20	0·21	0·21	0·22	0·22	0·22	0·23	0·23	0·24	0·24	0·24	0·25	0·25	0·26	0·27
8·1	0·16	0·16	0·16	0·16	0·17	0·17	0·17	0·17	0·18	0·18	0·18	0·19	0·19	0·20	0·20	0·20
8·2	0·12	0·12	0·12	0·13	0·13	0·13	0·13	0·13	0·13	0·14	0·14	0·14	0·14	0·15	0·15	0·15
8·3	0·09	0·09	0·09	0·10	0·10	0·10	0·10	0·10	0·10	0·10	0·11	0·11	0·11	0·11	0·11	0·11
8·4	0·07	0·07	0·07	0·07	0·07	0·07	0·07	0·08	0·08	0·08	0·08	0·08	0·08	0·08	0·08	0·09
8·5	0·05	0·05	0·05	0·05	0·05	0·05	0·05	0·06	0·06	0·06	0·06	0·06	0·06	0·06	0·06	0·06

171

TABLE 25 (cont.)

Cl = 19‰

°C / pH	0	2	4	6	8	10	12	14	16	18	20	22	24	26	28	30
7·4	0·82	0·84	0·86	0·88	0·90	0·92	0·94	0·96	0·99	1·01	1·03	1·06	1·08	1·11	1·14	1·18
7·5	0·65	0·66	0·68	0·69	0·71	0·72	0·74	0·76	0·77	0·79	0·81	0·83	0·85	0·87	0·90	0·97
7·6	0·51	0·52	0·53	0·55	0·56	0·57	0·58	0·59	0·61	0·62	0·63	0·64	0·66	0·68	0·70	0·72
7·7	0·40	0·41	0·42	0·43	0·44	0·45	0·46	0·47	0·48	0·49	0·50	0·51	0·52	0·53	0·55	0·56
7·8	0·31	0·32	0·33	0·33	0·34	0·35	0·36	0·36	0·37	0·36	0·38	0·39	0·41	0·41	0·42	0·43
7·9	0·24	0·25	0·25	0·26	0·26	0·27	0·28	0·28	0·29	0·29	0·30	0·30	0·31	0·32	0·32	0·33
8·0	0·19	0·20	0·20	0·20	0·20	0·21	0·21	0·22	0·22	0·22	0·23	0·23	0·24	0·24	0·25	0·25
8·1	0·15	0·15	0·15	0·15	0·16	0·16	0·16	0·16	0·17	0·17	0·17	0·18	0·18	0·18	0·19	0·19
8·2	0·11	0·12	0·12	0·12	0·12	0·12	0·12	0·13	0·13	0·13	0·13	0·13	0·14	0·14	0·14	0·15
8·3	0·09	0·09	0·09	0·09	0·09	0·09	0·09	0·09	0·10	0·10	0·10	0·10	0·10	0·10	0·11	0·11
8·4	0·07	0·07	0·07	0·07	0·07	0·07	0·07	0·07	0·07	0·07	0·07	0·07	0·07	0·07	0·08	0·08
8·5	0·05	0·05	0·05	0·05	0·05	0·05	0·05	0·05	0·05	0·05	0·05	0·05	0·05	0·05	0·06	0·06

Cl = 21‰

°C / pH	0	2	4	6	8	10	12	14	16	18	20	22	24	26	28	30
7·4	0·79	0·81	0·83	0·85	0·87	0·89	0·92	0·93	0·95	0·98	1·00	1·02	1·05	1·07	1·11	1·14
7·5	0·63	0·64	0·66	0·67	0·69	0·70	0·72	0·73	0·75	0·77	0·78	0·80	0·82	0·84	0·87	0·89
7·6	0·49	0·50	0·52	0·53	0·54	0·55	0·56	0·57	0·58	0·60	0·61	0·62	0·64	0·65	0·67	0·69
7·7	0·39	0·40	0·40	0·41	0·42	0·43	0·44	0·45	0·46	0·47	0·48	0·49	0·50	0·51	0·52	0·54
7·8	0·30	0·31	0·31	0·32	0·33	0·33	0·34	0·35	0·35	0·36	0·37	0·37	0·38	0·39	0·40	0·42
7·9	0·23	0·24	0·24	0·25	0·25	0·26	0·26	0·27	0·27	0·28	0·28	0·29	0·29	0·30	0·31	0·32
8·0	0·18	0·18	0·19	0·19	0·19	0·20	0·20	0·21	0·21	0·21	0·22	0·22	0·22	0·23	0·24	0·24
8·1	0·14	0·14	0·14	0·15	0·15	0·15	0·15	0·16	0·16	0·16	0·16	0·17	0·17	0·18	0·18	0·18
8·2	0·11	0·11	0·11	0·11	0·11	0·11	0·12	0·12	0·12	0·12	0·12	0·12	0·13	0·13	0·13	0·13
8·3	0·08	0·08	0·08	0·09	0·09	0·09	0·09	0·09	0·09	0·09	0·09	0·09	0·09	0·10	0·10	0·10
8·4	0·06	0·06	0·06	0·06	0·06	0·07	0·07	0·07	0·07	0·07	0·07	0·07	0·07	0·07	0·07	0·07
8·5	0·05	0·05	0·05	0·05	0·05	0·05	0·05	0·05	0·05	0·05	0·05	0·05	0·05	0·05	0·05	0·05

Hence
$$K_1' = \frac{a_{H^.} \times C_{HCO_3'}}{P_{CO_2} \times \alpha_0 \times a_{H_2O}}$$

From the relation

Carbonate alkalinity $= C_{HCO_3'} + 2C_{CO_3''}$, and the definition of K_2',

P_{CO_2} = carbonate alkalinity $\times \dfrac{a_{H^.}}{K_1'\alpha_0 \left(1 + \dfrac{2K_2'}{a_{H^.}}\right) a_{H_2O}}$ atmospheres.

(Equation 2)

Values of the term $\dfrac{a_{H^.}}{K_1'\alpha_0 \left(1 + \dfrac{2K_2'}{a_{H^.}}\right) a_{H_2O}}$ are given for sea waters

over a range of salinities, temperatures and hydrogen-ion concentrations in table 25.

From the relation

$$\text{Total } CO_2 = C_{HCO_3'} + C_{CO_3''} + C_{CO_2} \text{ in solution,}$$

since $C_{CO_2} = \alpha_s P_{CO_2}$, where α_s is the concentration of CO_2 in mols per litre soluble at 1 atmosphere in sea water of the particular salinity and temperature under consideration (table 23):

Total CO_2 = carbonate alkalinity $\times \dfrac{1 + \dfrac{K_2'}{a_{H^.}} + \dfrac{\alpha_s \times a_{H^.}}{K_1' \times \alpha_0 \times a_{H_2O}}}{1 + \dfrac{2K_2'}{a_{H^.}}}$

g.mols per litre.

(Equation 3)

Values of the final term in this equation are given for sea waters over a range of salinities, temperatures and hydrogen-ion concentrations in table 26 (Buch 1951).

From table 20 the carbonate alkalinity of a water can be quickly calculated, from table 25 its partial pressure of carbon dioxide, P_{CO_2}, and from table 26 its total content of carbon dioxide, ΣCO_2.

From the following equations, derived from the definition of carbonate alkalinity and of K_2', the concentrations of bicarbonate, of carbonate and of molecular CO_2 in solution can be calculated:

$$C_{HCO_3'} = \text{carbonate alkalinity} \times \frac{1}{1 + \dfrac{2K_2'}{a_H}}.$$

(Equation 4)

$$C_{CO_3'} = \text{carbonate alkalinity} \times \frac{K_2'}{2K_2' + a_H}.$$

(Equation 5)

$$C_{CO_2} = P_{CO_2} \times \alpha_s.$$ (Equation 6)

USE OF THE TABLES AND CONSTANTS

1. What is the total CO_2 content of a sea water of $35\cdot35\%_0$ S, having a hydrogen-ion concentration of pH $= 8\cdot3$ when at $16°$ C.? From fig. 58, $35\cdot35\%_0$ S $= 19\cdot5\%_0$ Cl. If the water is a natural water to which no acid or alkali has been added, its titration alkalinity $= 12\cdot3 \times 19\cdot5 \times 10^{-5}$ (p. 162), otherwise it is necessary to determine the titration alkalinity by the method described, which is that which has been used in the experimental determination of the constants shown in the several tables.

From table 20,

Carbonate alkalinity $= 240 - 11$,

or 229×10^{-5} normal.

From table 26,

Total CO_2 at $16°$ C. and $8\cdot3$ pH $= 229 \times 0\cdot87 \times 10^{-5}$,

or 199×10^{-5} g.mols per litre

or $44\cdot5$ cm.^{3} per litre.

2. What is the concentration of molecular CO_2 in solution (C_{CO_2}) in a sea water of $19\cdot5\%_0$ Cl, titration alkalinity of 244×10^{-5} normal, at $16°$ C. and pH $7\cdot4$?

From table 25,

P_{CO_2} at $16°$ and $7\cdot4$ pH $= 242 \times 10 \times 0\cdot98$,

or 237×10 atmospheres.

From table 23, $\alpha_s = 379 \times 10^{-4}$.

TABLE 26. *Factors which when multiplied by the carbonate alkalinity yield the value of the total carbon dioxide in mols per litre*

Cl = 15‰

°C. pH	0	2	4	6	8	10	12	14	16	18	20	22	24	26	28	30
7·4	1·05	1·05	1·04	1·04	1·04	1·03	1·03	1·03	1·03	1·02	1·02	1·02	1·02	1·01	1·01	1·01
7·5	1·03	1·03	1·03	1·02	1·02	1·02	1·02	1·01	1·01	1·01	1·01	1·00	1·00	1·00	1·00	1·00
7·6	1·02	1·02	1·01	1·01	1·01	1·01	1·00	1·00	1·00	1·00	1·00	0·99	0·99	0·99	0·99	0·99
7·7	1·01	1·01	1·00	1·00	1·00	1·00	0·99	0·99	0·99	0·99	0·98	0·98	0·98	0·98	0·97	0·97
7·8	1·00	0·99	0·99	0·99	0·99	0·98	0·98	0·98	0·98	0·97	0·97	0·97	0·97	0·96	0·96	0·96
7·9	0·99	0·98	0·98	0·98	0·97	0·97	0·97	0·97	0·96	0·96	0·96	0·96	0·95	0·95	0·95	0·95
8·0	0·98	0·97	0·97	0·96	0·96	0·96	0·96	0·95	0·95	0·95	0·94	0·94	0·94	0·93	0·93	0·93
8·1	0·96	0·96	0·95	0·95	0·95	0·94	0·94	0·94	0·93	0·93	0·93	0·92	0·92	0·92	0·91	0·91
8·2	0·95	0·95	0·94	0·94	0·93	0·93	0·92	0·92	0·92	0·91	0·91	0·90	0·90	0·90	0·89	0·89
8·3	0·93	0·93	0·92	0·92	0·91	0·91	0·90	0·90	0·90	0·89	0·89	0·88	0·88	0·88	0·87	0·87
8·4	0·92	0·91	0·90	0·90	0·89	0·89	0·88	0·88	0·87	0·87	0·86	0·86	0·86	0·85	0·85	0·84
8·5	0·90	0·89	0·88	0·88	0·87	0·87	0·86	0·86	0·85	0·85	0·84	0·84	0·83	0·83	0·82	0·82

Cl = 17‰

°C. pH	0	2	4	6	8	10	12	14	16	18	20	22	24	26	28	30
7·4	1·04	1·04	1·04	1·03	1·03	1·03	1·03	1·02	1·02	1·02	1·02	1·02	1·01	1·01	1·01	1·01
7·5	1·03	1·03	1·02	1·02	1·02	1·02	1·01	1·01	1·01	1·01	1·00	1·00	1·00	1·00	1·00	0·99
7·6	1·02	1·02	1·01	1·01	1·01	1·01	1·00	1·00	1·00	0·99	0·99	0·99	0·99	0·99	0·98	0·98
7·7	1·01	1·00	1·00	1·00	1·00	0·99	0·99	0·99	0·99	0·98	0·98	0·98	0·98	0·97	0·97	0·97
7·8	1·00	0·99	0·99	0·99	0·98	0·98	0·98	0·98	0·97	0·97	0·97	0·96	0·96	0·96	0·96	0·95
7·9	0·99	0·98	0·98	0·98	0·97	0·97	0·96	0·96	0·96	0·95	0·95	0·95	0·95	0·94	0·94	0·94
8·0	0·97	0·97	0·96	0·96	0·96	0·95	0·95	0·95	0·94	0·94	0·93	0·93	0·93	0·93	0·92	0·92
8·1	0·96	0·95	0·95	0·95	0·94	0·94	0·93	0·93	0·92	0·92	0·92	0·91	0·91	0·91	0·90	0·90
8·2	0·94	0·93	0·93	0·92	0·92	0·91	0·91	0·91	0·90	0·90	0·90	0·90	0·89	0·89	0·88	0·88
8·3	0·93	0·92	0·92	0·91	0·91	0·90	0·89	0·89	0·89	0·88	0·88	0·87	0·87	0·86	0·86	0·85
8·4	0·91	0·90	8·87	0·89	0·88	0·88	0·87	0·87	0·86	0·86	0·85	0·85	0·84	0·84	0·83	0·83
8·5	0·89	0·88	0·87	0·87	0·86	0·85	0·85	0·84	0·84	0·83	0·82	0·82	0·82	0·81	0·81	0·80

TABLE 26 (cont.)

Cl = 19‰

°C. pH	0	2	4	6	8	10	12	14	16	18	20	22	24	26	28	30
7·4	1·04	1·04	1·03	1·03	1·03	1·02	1·02	1·02	1·02	1·01	1·01	1·01	1·01	1·01	1·00	1·00
7·5	1·03	1·02	1·02	1·02	1·01	1·01	1·01	1·01	1·00	1·00	1·00	1·00	0·99	0·99	0·99	0·99
7·6	1·01	1·01	1·01	1·00	1·00	1·00	1·00	0·99	0·99	0·99	0·99	0·98	0·98	0·98	0·98	0·98
7·7	1·00	1·00	1·00	0·99	0·99	0·99	0·98	0·98	0·98	0·98	0·97	0·97	0·97	0·97	0·96	0·96
7·8	0·99	0·99	0·98	0·98	0·98	0·97	0·97	0·97	0·96	0·96	0·96	0·96	0·95	0·95	0·95	0·95
7·9	0·98	0·97	0·97	0·97	0·96	0·96	0·96	0·95	0·95	0·95	0·94	0·94	0·94	0·93	0·93	0·93
8·0	0·96	0·96	0·95	0·95	0·95	0·94	0·94	0·94	0·93	0·93	0·93	0·92	0·92	0·92	0·91	0·91
8·1	0·95	0·94	0·94	0·94	0·93	0·93	0·92	0·92	0·92	0·91	0·91	0·90	0·90	0·90	0·89	0·89
8·2	0·93	0·93	0·92	0·92	0·92	0·91	0·91	0·90	0·90	0·89	0·89	0·88	0·88	0·88	0·87	0·87
8·3	0·92	0·91	0·91	0·90	0·90	0·89	0·88	0·88	0·87	0·87	0·87	0·86	0·86	0·85	0·85	0·84
8·4	0·90	0·89	0·89	0·88	0·87	0·87	0·86	0·86	0·85	0·84	0·84	0·83	0·83	0·82	0·82	0·81
8·5	0·87	0·87	0·86	0·86	0·85	0·84	0·84	0·83	0·83	0·82	0·81	0·81	0·80	0·80	0·79	0·79

Cl = 21‰

°C. pH	0	2	4	6	8	10	12	14	16	18	20	22	24	26	28	30
7·4	1·03	1·03	1·03	1·03	1·02	1·02	1·02	1·02	1·01	1·01	1·01	1·01	1·00	1·00	1·00	1·00
7·5	1·02	1·02	1·02	1·01	1·01	1·01	1·01	1·00	1·00	1·00	1·00	0·99	0·99	0·99	0·99	0·98
7·6	1·01	1·00	1·00	1·00	1·00	0·99	0·99	0·99	0·99	0·98	0·98	0·98	0·98	0·97	0·97	0·97
7·7	0·99	0·99	0·99	0·99	0·98	0·98	0·98	0·97	0·97	0·97	0·97	0·96	0·96	0·96	0·96	0·96
7·8	0·98	0·98	0·98	0·97	0·97	0·97	0·96	0·96	0·96	0·96	0·95	0·95	0·95	0·95	0·94	0·94
7·9	0·97	0·97	0·96	0·96	0·96	0·95	0·95	0·95	0·94	0·94	0·94	0·93	0·93	0·93	0·92	0·92
8·0	0·96	0·95	0·95	0·95	0·94	0·94	0·93	0·93	0·93	0·92	0·92	0·92	0·91	0·91	0·91	0·90
8·1	0·94	0·94	0·93	0·93	0·92	0·92	0·92	0·91	0·91	0·90	0·90	0·89	0·89	0·89	0·88	0·88
8·2	0·93	0·92	0·92	0·91	0·91	0·90	0·90	0·89	0·89	0·88	0·88	0·87	0·87	0·86	0·86	0·85
8·3	0·91	0·90	0·90	0·89	0·89	0·88	0·87	0·87	0·86	0·86	0·85	0·85	0·84	0·84	0·83	0·83
8·4	0·89	0·88	0·87	0·87	0·86	0·86	0·85	0·85	0·84	0·83	0·83	0·82	0·82	0·81	0·81	0·80
8·5	0·86	0·86	0·85	0·84	0·84	0·83	0·82	0·82	0·81	0·81	0·80	0·79	0·79	0·78	0·78	0·77

From the relation $C_{CO_2} = P_{CO_2} \times \alpha_s$

$$C_{CO_2} = 94 \times 10^{-6} \text{ g.mols per litre,}$$
or $2 \cdot 1$ cm.^{3} per litre.

3. What is the concentration of carbonate ions, $C_{CO_3''}$ in a sea water of $19 \cdot 5 \%_0$ Cl at $10°$ C., pH $8 \cdot 5$, and titration alkalinity 244×10^{-5} normal?

TABLE 27. *Conversion of* pH *to* $C_{H^.}$ *or* $a_{H^.}$*
in g. per litre

pH	C_H or a_H	pH	$C_{H^.}$ or $a_{H^.}$	pH	C_H or $a_{H^.}$
Q 0·00	$1 \cdot 000 \times 10^{-Q}$	Q 0·34	$0 \cdot 457 \times 10^{-Q}$	Q 0·67	$0 \cdot 214 \times 10^{-Q}$
0·01	0·977	0·35	0·447	0·68	0·209
0·02	0·955	0·36	0·437	0·69	0·204
0·03	0·933	0·37	0·427	0·70	0·200
0·04	0·912	0·38	0·417	0·71	0·195
0·05	0·891	0·39	0·407	0·72	0·191
0·06	0·871	0·40	0·398	0·73	0·186
0·07	0·851	0·41	0·389	0·74	0·182
0·08	0·832	0·42	0·380	0·75	0·178
0·09	0·813	0·43	0·372	0·76	0·174
0·10	0·794	0·44	0·363	0·77	0·170
0·11	0·776	0·45	0·355	0·78	0·166
0·12	0·759	0·46	0·347	0·79	0·162
0·13	0·741	0·47	0·339	0·80	0·158
0·14	0·725	0·48	0·331	0·81	0·155
0·15	0·709	0·49	0·324	0·82	0·151
0·16	0·692	0·50	0·316	0·83	0·148
0·17	0·676	0·51	0·309	0·84	0·144
0·18	0·661	0·52	0·302	0·85	0·141
0·19	0·646	0·53	0·295	0·86	0·138
0·20	0·631	0·54	0·288	0·87	0·135
0·21	0·617	0·55	0·282	0·88	0·132
0·22	0·603	0·56	0·275	0·89 ·	0·129
0·23	0·589	0·57	0·269	0·90	0·126
0·24	0·575	0·58	0·263	0·91	0·123
0·25	0·562	0·59	0·257	0·92	0·120
0·26	0·549	0·60	0·251	0·93	0·117
0·27	0·537	0·61	0·245	0·94	0·115
0·28	0·525	0·62	0·240	0·95	0·112
0·29	0·513	0·63	0·234	0·96	0·110
0·30	0·501	0·64	0·229	0·97	0·107
0·31	0·490	0·65	0·224	0·98	0·105
0·32	0·479	0·66	0·219	0·99	0·102
0·33	0·468				

* For working purposes C_H is taken to equal a_H.

From table 20,

Carbonate alkalinity at $10°$ C. and $8\cdot5$ pH $= 244-14$,

or 230×10^{-5} normal.

From table 27,

pH $8\cdot5 = 0\cdot316 \times 10^{-8}$ g.ions per litre $= a_H$..

From table 22, K_2' at $10°$ C. $= 0\cdot76 \times 10^{-9}$.

From equation (5)

$$C_{CO_3''} = \text{carbonate alkalinity} \times \frac{K_2'}{2K_2' + a_H} \ ;$$

on substituting the values shown above

$$C_{CO_3''} = 37\cdot3 \times 10^{-5} \text{ g.mols per litre,}$$
or $0\cdot000746$ equivalents per litre.

4. At what pH is a sea water of salinity $27\%_0$ in equilibrium with an inert gas containing 5% CO_2 at $14°$ C.?

$$27\%_0 \ S = 15\%_0 \ Cl \quad (\text{fig. 58}).$$
$$P_{CO_2} = 0\cdot05 \text{ atmosphere.}$$

Since no carbonate or borate ions exist at a hydrogen-ion concentration lower than about pH $7\cdot2$, 'Carbonate alkalinity' = 'titration' alkalinity and from the relation 'Titration alkalinity' $= 12\cdot3 \times Cl\%_0 \times 10^{-5}$, when no acid or alkali has been added to the water:

'Carbonate alkalinity' $= 185 \times 10^{-5}$.

From table 21, K_1' at $14°$ C. $= 0\cdot80 \times 10^{-6}$.

From table 22, K_2' at $14°$ C. $= 0\cdot66 \times 10^{-9}$.

From table 23, $\quad \alpha_0 = 472 \times 10^{-4}$.

From table 24, $\quad a_{H_2O} = 0\cdot985$.

On substituting these values in equation (2) and solving, $a_H = 10^{-6}$ g. per litre, or pH $= 6\cdot0$.

EXPERIMENTAL DETERMINATION OF THE APPARENT
DISSOCIATION CONSTANTS

A brief description of the methods employed to determine the
two CO_2 dissociation constants makes clearer their meaning and
their use.

The first apparent constant K_1' alone shows the relation
between pH, bicarbonate ions and CO_2 in solution in sea water
with a pH below about 7, when carbonate ions almost cease to
exist. In order to determine this constant, sea water was shaken
for about 10 minutes at constant temperature with air enriched
with CO_2, the air in contact with the sample of sea water being
kept at atmospheric pressure. Equilibrium being attained, the
partial pressure of CO_2 in the air in contact with the water, and
hence in the water, was found by determining the CO_2 content
of this air gasometrically. This provides P_{CO_2} in equation (2).
The pH (pa_H·) of the water was determined with a quinhydrone
electrode. The term $2K_2'/a_H$ in the equations becomes negligibly
small with a pH below 7. Hence, having determined the excess
base and calculated the carbonate alkalinity, K_1' can be found
from the equation. From a number of such calculations for a
particular sea water in equilibrium with air of varying CO_2
content, its K_1' was obtained. Similar determinations at
other temperatures, and for waters of differing salinities,
gave the variation of K_1' with temperature and salinity.
In practice deductions were made from the observation of
other workers concerning the effect of temperature and
salt concentration on the dissociation of sodium bicarbonate
solutions.

The practical determination of the second constant, K_2',
presented greater difficulties. It was necessary to work at
hydrogen-ion concentrations where carbonate ions formed a
material proportion of the bound CO_2. Under these conditions,
P_{CO_2} is very small and the experimental error in its determination affects the result. In order to overcome the first difficulty,
the total CO_2 has been used. This is determined by boiling off
the dissolved gases from the water sample under reduced
pressure after adding phosphoric acid. The pH was measured
colorimetrically in the earlier experiments (Buch, Harvey,

Wattenberg and Gripenberg, 1932; Moberg, Greenberg, Revelle and Allen, 1934) and later with a hydrogen electrode (Buch, 1938). The water in these experiments was brought to a pH where the carbonate exceeded the bicarbonate ions (c. pH 8·8) by aerating with CO_2 free air or with hydrogen, and in the latter experiments an artificial sea water without boric acid was used in order to omit this further complication. It is to be noted that in this relatively high range of pH, the excess of hydroxyl ions over hydrogen ions is material, and the last term in equation (1) cannot be neglected.

Moberg et al. employed a titration method for estimating the total CO_2 in the water. The quantity of strong acid required to bring a sea water to the turning point of phenolphthalein (at which point almost all the carbonate ions originally present have changed to bicarbonate ions and free CO_2) was taken as strictly equivalent to the carbonate originally present. The effect of the borate ions was neglected: In waters of high pH, the quantity of acid required by the hydroxyl ions was subtracted. The equivalence of acid required was taken as equal to $C_{CO_3''}$ and the excess base as $C_{HCO_3'} + 2C_{CO_3''}$. The difference (that is, $C_{HCO_3'} + C_{CO_3''}$) was taken as equal to the total CO_2, since the quantity of free CO_2 in solution above pH 8·1 is only a fraction of 1 % of the combined CO_2. The rough values for CO_2 obtained in this way lay within $3\frac{1}{2}$ ‰ of the values found by a manometric technique. The titration method is discussed by Greenberg, Moberg and Allen (1932).

The Effect of Pressure at Great Depths

With increase in pressure there is a change in the CO_2 system. The two constants K_1' and K_2' both increase. The water is compressed; the pressure at 10,000 m. depth causes a diminution of some 4 % in volume. In consequence both the 'excess base' and total CO_2, which are expressed in terms of equivalents and gramm-mols per litre respectively, are increased.

As a result of pressure, the free unbound CO_2 in solution decreases, the bicarbonate increases and the hydrogen-ion concentration increases (p. 156) (Buch and Gripenberg, 1932).

The following table, from Buch and Gripenberg, shows the changes due to increasing depth for a water of $Cl\%_{00} = 19{\cdot}5$ at $0°$ C.:

Depth in metres	pH	Excess base	$C_{\Sigma CO_2}$	C_{CO_2}	$C_{HCO_3'}$	$C_{CO_3''}$
0	8·00	2·40	2·23	0·026	2·00	0·20
5,000	7·89	2·45	2·27	0·020	2·05	0·20
10,000	7·78	2·50	2·31	0·015	2·09	0·20
0	7·60	2·40	2·38	0·074	2·22	0·09
5,000	7·44	2·45	2·42	0·063	2·29	0·08
10,000	7·27	2·50	2·48	0·052	2·36	0·07

The effect of pressure on the first dissociation constant has been investigated by Brander. The effect on the second constant K_2' is considered to be the same as the effect of pressure on other weak acids; presumably K_B' is affected in the same way. The following values have been taken from the table compiled by Wattenberg (1933, p. 251):

Depth in metres	Percentage increase in K_1'	Percentage increase in K_2'
0		
2,000	26	9·6
4,000	58	20
6,000	100	32
10,000	202	55

INTERCHANGE OF CARBON DIOXIDE BETWEEN THE SEA AND THE ATMOSPHERE

The CO_2 content of the atmosphere over the North Atlantic, North Sea and English Channel amounts to $0{\cdot}032$–$0{\cdot}033\%$ (32–33×10^{-5} atmosphere) and has increased from about $0{\cdot}030\%$ during the last fifty years. This increase, shown by numerous analyses by independent observers using different methods, conforms with the addition made to the atmosphere by industrial combustion of coal and oil over this period (Buch, 1952). Observations have shown that over the Arctic the CO_2 in the atmosphere is slightly less (Buch, 1939a, b).

Krogh (1910) determined the partial pressure of CO_2 in a series of samples from the surface between Scotland and Canada by shaking the water with a small bubble of air. After shaking, such a bubble contained c. 0·03 % of CO_2, indicating a partial pressure of CO_2 amounting to 30×10^{-5} atmosphere. In general, the partial pressure in the water was less than the partial pressure in the atmosphere. Hence it is absorbed from the air, the sea acting as a regulator of the CO_2 in the atmosphere. Buch (1939a, b) has made similar observations during summer months between Denmark and America and between Norway and Iceland. The partial pressure in the surface water was found experimentally and by calculation from pH, salinity and temperature. It varied, but on an average was some 5 % less than in the atmosphere. In one series of observations the partial pressure was found to increase with height above the sea surface. These series of observations confirmed Krogh's conclusions.

During the first half of the yeàr, the CO_2 content of the upper layer decreases due to abstraction by the plants during daylight exceeding replacement by respiration of animals, bacteria and of plants. As a result of predominant utilization by plants, the partial pressure of CO_2 in the surface water of the sea decreases, while during the same period the temperature of the water rises, which has the effect of increasing the partial pressure of CO_2 in the surface water. The two factors—utilization by plants and rise in temperature—act in opposition to each other.

A series of observations in the English Channel, in which pH and titration alkalinity were determined at intervals throughout a year, show the combined effect of biological and temperature change in the surface water (fig. 65).

A marked decrease in partial pressure occurred between mid-February and mid-May during the period of maximum phytoplankton population, which utilized CO_2 and thereby increased the pH of the water.

Deacon (1940) gives data for the waters of the subtropical South Atlantic and of the Antarctic, which also show a greater partial pressure during the winter months.

In a general review, Buch (1952) assesses the partial pressure in the surface water as varying around 29×10^{-5} in temperate regions and 15×10^{-5} in Arctic waters, indicating absorption of

CO_2 from the atmosphere, while in tropical latitudes the partial pressure may be as high as, or even higher than, that in the atmosphere.

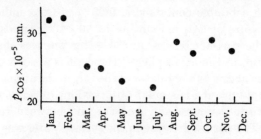

FIG. 65. The seasonal change in partial pressure of CO_2 in the surface water of the English Channel, in atmospheres $\times 10^{-5}$. (After Cooper, 1933.)

From crude observations the rate of absorption of CO_2 by sea water appears to be notably slow. This may be due to the slow rate at which molecular CO_2 changes into H_2CO_3 prior to ionization, whereas in living tissue this rate is much greater due to the catalytic action of carbonic anhydrase.

183

CHAPTER XI

DISSOLVED OXYGEN, NITROGEN AND INERT GASES

DISSOLVED OXYGEN

THE oxygen content, of waters of varying salinity when saturated with air, has been determined by Fox (1907) between 0 and 30° C., and is shown in table 28. These values appear to be about 3% too high (Truesdale and Downing, 1954, *Nature* **173**, ⋆ 1236; Allen, 1955, *Nature* **175**, 83).

In ordinary practice the oxygen in sea water is estimated by Winkler's method (Jacobsen and Knudsen, 1921). In waters of low oxygen content difficulty has been found in obtaining a precise end-point with starch as indicator, and has been overcome by a simple amperometric titration (Knowles and Lowden, 1953).

Where only small quantities of water are available, as in some physiological experiments, a modification of this method, due to Krogh, gives an accuracy of 2 %, using 1–2 cm³ of sea water (Fox and Wingfield, 1938).

The rate at which oxygen enters undersaturated water from the atmosphere has received attention. This rate shows a direct relation to the degree of undersaturation and is controlled by the rapidity with which the water at the air-water interface is renewed. Krogh (1910) measured the rate of solution of oxygen by water flowing past a small bubble of the gas. He observed that lower rates were found when using larger bubbles, and attributed this to incomplete renovation of the surface. Adeney (1928) has measured the rate of solution of oxygen from a bubble of air caused to travel up and down a tube filled with fresh and sea waters at various temperatures. Although deaerated fresh water absorbs oxygen more rapidly than sea water, it required the same time for either to attain the same percentage saturation.

A relation between time, percentage saturation and temperature has been derived (Adeney, p. 68) from the data obtained by this method. From this, the *rate of invasion* of oxygen from air is deduced:

Oxygen in cm³ entering 1 cm² per minute
$$= 9{\cdot}6(t° + 36)\,(a - x)\,10^{-6},$$

TABLE 28. *Number of cubic centimetres of oxygen at N.T.P. dissolved in 1 litre of sea water saturated with air at the temperature shown and at 760 mm. pressure*

The chlorine content of the sea water is given in grams Cl per 1000 g. of sea water (Fox).

°C.	Cl=0	Cl=1	Cl=2	Cl=3	Cl=4	Cl=5	Cl=6	Cl=7	Cl=8	Cl=9	Cl=10	Cl=11	Cl=12	Cl=13	Cl=14	Cl=15	Cl=16	Cl=17	Cl=18	Cl=19	Cl=20
0	10·29	10·17	10·06	9·94	9·83	9·71	9·59	9·48	9·36	9·25	9·13	9·01	8·90	8·78	8·66	8·55	8·43	8·32	8·20	8·08	7·97
1	10·02	9·90	9·79	9·68	9·57	9·45	9·34	9·23	9·12	9·01	8·89	8·78	8·67	8·56	8·44	8·33	8·22	8·11	8·00	7·88	7·77
2	9·75	9·64	9·53	9·43	9·32	9·21	9·10	8·99	8·88	8·75	8·67	8·56	8·45	8·34	8·23	8·12	8·02	7·91	7·80	7·69	7·58
3	9·50	9·39	9·29	9·19	9·08	8·98	8·87	8·77	8·66	8·56	8·45	8·35	8·24	8·14	8·03	7·93	7·82	7·72	7·61	7·51	7·40
4	9·26	9·16	9·06	8·95	8·85	8·75	8·65	8·55	8·45	8·35	8·24	8·14	8·04	7·94	7·84	7·74	7·64	7·53	7·43	7·33	7·23
5	9·03	8·93	8·83	8·73	8·64	8·54	8·44	8·34	8·24	8·15	8·05	7·95	7·85	7·75	7·65	7·56	7·46	7·36	7·26	7·16	7·07
6	8·81	8·71	8·62	8·52	8·43	8·33	8·24	8·14	8·05	7·95	7·86	7·76	7·67	7·57	7·48	7·38	7·28	7·20	7·10	7·01	6·91
7	8·60	8·50	8·41	8·32	8·23	8·14	8·04	7·95	7·85	7·77	7·68	7·59	7·50	7·40	7·31	7·22	7·13	7·04	6·95	6·85	6·76
8	8·40	8·31	8·22	8·13	8·04	7·95	7·86	7·77	7·68	7·59	7·51	7·42	7·33	7·24	7·15	7·06	6·97	6·89	6·80	6·71	6·62
9	8·21	8·12	8·03	7·95	7·86	7·77	7·69	7·60	7·52	7·43	7·34	7·26	7·17	7·09	7·00	6·91	6·83	6·74	6·66	6·57	6·48
10	8·02	7·94	7·85	7·77	7·69	7·60	7·52	7·44	7·36	7·27	7·19	7·10	7·02	6·94	6·85	6·77	6·69	6·60	6·52	6·44	6·35
11	7·84	7·76	7·68	7·60	7·52	7·44	7·36	7·28	7·20	7·12	7·04	6·96	6·88	6·80	6·71	6·63	6·55	6·47	6·39	6·31	6·23
12	7·68	7·60	7·52	7·44	7·36	7·29	7·21	7·13	7·05	6·97	6·89	6·82	6·74	6·66	6·58	6·50	6·43	6·35	6·27	6·19	6·11
13	7·52	7·44	7·36	7·29	7·21	7·14	7·06	6·98	6·91	6·83	6·76	6·68	6·61	6·53	6·46	6·38	6·31	6·23	6·15	6·08	6·00
14	7·37	7·29	7·21	7·14	7·07	7·00	6·92	6·85	6·77	6·70	6·63	6·55	6·48	6·41	6·34	6·26	6·19	6·11	6·04	5·97	5·89
15	7·22	7·15	7·07	7·00	6·93	6·86	6·79	6·72	6·64	6·57	6·50	6·43	6·36	6·29	6·22	6·14	6·07	6·00	5·93	5·86	5·79
16	7·08	7·01	6·94	6·87	6·80	6·73	6·66	6·59	6·52	6·45	6·38	6·31	6·24	6·17	6·10	6·03	5·96	5·89	5·82	5·76	5·69
17	6·94	6·88	6·81	6·74	6·67	6·60	6·54	6·47	6·40	6·33	6·26	6·20	6·13	6·06	5·99	5·93	5·86	5·79	5·72	5·66	5·59
18	6·81	6·75	6·68	6·62	6·55	6·48	6·42	6·35	6·28	6·22	6·15	6·09	6·02	5·96	5·89	5·83	5·76	5·69	5·63	5·56	5·49
19	6·69	6·63	6·56	6·50	6·44	6·37	6·30	6·24	6·17	6·11	6·05	5·98	5·92	5·86	5·79	5·73	5·66	5·60	5·53	5·47	5·40
20	6·57	6·51	6·44	6·38	6·33	6·26	6·19	6·13	6·07	6·00	5·95	5·88	5·82	5·76	5·69	5·63	5·56	5·50	5·44	5·38	5·31
21	6·46	6·40	6·33	6·27	6·22	6·15	6·09	6·03	5·96	5·90	5·85	5·78	5·72	5·66	5·59	5·53	5·47	5·41	5·35	5·29	5·22
22	6·35	6·29	6·23	6·17	6·11	6·04	5·98	5·92	5·86	5·80	5·75	5·68	5·62	5·56	5·50	5·44	5·38	5·32	5·26	5·20	5·13
23	6·24	6·18	6·12	6·06	6·01	5·94	5·88	5·82	5·77	5·71	5·65	5·59	5·53	5·47	5·41	5·35	5·29	5·23	5·17	5·11	5·04
24	6·14	6·08	6·02	5·97	5·91	5·84	5·79	5·73	5·67	5·61	5·55	5·50	5·44	5·38	5·32	5·26	5·20	5·14	5·09	5·03	4·95
25	6·04	5·99	5·92	5·87	5·81	5·75	5·69	5·64	5·58	5·52	5·46	5·41	5·35	5·29	5·23	5·17	5·12	5·06	5·00	4·95	4·86
26	5·94	5·89	5·82	5·77	5·71	5·66	5·60	5·55	5·49	5·43	5·37	5·32	5·26	5·20	5·14	5·09	5·03	4·97	4·92	4·86	4·78
27	5·84	5·79	5·73	5·67	5·62	5·57	5·51	5·46	5·40	5·34	5·28	5·23	5·17	5·11	5·06	5·00	4·94	4·89	4·83	4·78	4·70
28	5·75	5·69	5·64	5·58	5·53	5·48	5·42	5·37	5·31	5·25	5·19	5·14	5·08	5·02	4·97	4·91	4·86	4·80	4·75	4·69	4·62
29	5·66	5·60	5·55	5·49	5·44	5·39	5·33	5·28	5·22	5·16	5·10	5·05	4·99	4·93	4·88	4·83	4·77	4·71	4·66	4·60	4·54
30	5·57	5·51	5·46	5·40	5·35	5·30	5·24	5·19	5·13	5·07	5·01	4·96	4·90	4·85	4·79	4·74	4·68	4·63	4·58	4·52	4·46

where $t°$ is the temperature of the water, x cm.³ per litre is the concentration of oxygen present in the water, a cm.³ per litre is the concentration of oxygen in the water at the temperature and salinity in question when saturated with air (table 1).

The rates found under these conditions of surface renovation are considered by Adeney to represent maximum values likely to occur under natural conditions, and to be about twice the average rates occurring under open-sea conditions.

The renovation of the surface, and consequently the rate of invasion, depends upon the extent to which the surface water is stirred by the wind. It is increased when the air is dry. Evaporation causes cooling at the surface and sets up convection currents which renovate the surface. This effect was found to be enhanced by increasing salinity up to about 15 S‰, evaporation causing an increase in density by concentration of the salts in addition to the increase in density caused by cooling.

Under quiescent conditions, as when deaerated water is exposed to the air in a vessel in the laboratory, the rate of invasion of oxygen may be less than one hundredth that when the water is kept moving and the surface renovated. Under quiescent conditions the moisture content of the air plays a major part in controlling the rate of invasion.

The *rate of liberation* of oxygen from supersaturated water does not appear to have been investigated experimentally; evidence of seasonal liberation from the surface in temperate latitudes is summarized on p. 35.

The distribution of oxygen in the ocean and the factors causing changes are discussed on pp. 32–36.

DISSOLVED NITROGEN AND INERT GASES

The nitrogen in solution in sea water of varying salinity when in equilibrium with the air has been found and tabulated over a range of temperature by Fox (1907). It was noted by Rakestraw, Herrick and Urry (1939) that the sea was slightly undersaturated with nitrogen more frequently than might be expected when they based the saturation on the values given in these tables; this led them to suspect that slightly supersaturated waters had been used in Fox's determinations. It is notable that, when sea water is shaken with air, it becomes supersaturated, to the extent of about 2 % after all bubbles have disappeared. Rakestraw and Emmel (1938) redetermined the solubility and obtained values slightly less than those previously found.

The solubility of both oxygen and nitrogen is affected similarly by changes in temperature and chlorinity. From the solubility values for oxygen (p. 184) the values for nitrogen may be calculated from the relation

$$\text{cm}^3 \text{ N}_2 = \frac{\text{cm}^3 \text{ O}_2 + 0 \cdot 22}{0 \cdot 577},$$

within the range of 15–22‰ Cl, where cm^3 N denotes the volume of nitrogen, exclusive of inert gases, per litre, measured at N.T.P.

The *inert* gases, argon, etc., in solution are equal to about 2·7 % of the dissolved nitrogen. From Rakestraw and Emmel's data the effect of change of temperature and chlorinity on their solubility differs from the effect on nitrogen. They have also found evidence that traces of hydrogen or hydrocarbons occur in solution in inshore waters. The quantities of these inert gases in the sea appear to vary with the quantity of nitrogen in the water. Sea water also contains traces of helium and neon in solution; from 1·2 to 1·8 × 10⁻⁴ cm³ per litre of these gases were found by Rakestraw, Herrick and Urry (1939) in ocean waters from various depths.

The distribution of dissolved nitrogen in the oceans has received some attention, because the quantity in samples from various depths may give an indication of the temperature of the water when the water mass was in contact with the atmosphere.

It is considered to be a conservative constituent of the water, unaltered by biological changes. This is uncertain, since bacteria capable of forming nitrogen and bacteria capable of fixing atmospheric nitrogen are found in the sea (p. 78). However, the balance of evidence at present suggests that neither formation nor utilization takes place to any appreciable extent in the oceans.

The earlier data on nitrogen content have been reviewed by Buch (1929) and more recent data have been obtained by Rakestraw and Emmel at several deep-water positions in the Atlantic. The quantity of dissolved nitrogen found in the waters depended upon their temperature. It lay near the calculated requirement to saturate it, with respect to air, at the temperature of the water sample when brought to the surface. In the warmer upper layers supersaturation, up to about 8 %, was frequently met with. In this connexion it is of interest that if a saturated warm water is mixed with a saturated colder water, the resulting mixture will be supersaturated. Samples which were undersaturated, to a maximum extent of 5 %, were also found. The bottom water, at 3000–4500 m. depth, having a temperature of $3-2\frac{1}{2}^\circ$ C., was within ± 4 % of its saturation value, suggesting that it had not altered in temperature by more than $1\frac{1}{2}^\circ$ C. since it had been in contact with the atmosphere at the surface in polar regions.

CHAPTER XII

SOME ANALYTICAL TECHNIQUES

In collaboration with F. A. J. Armstrong

THE estimation of microconstituents which are available to plant life, when at the very great dilution which may occur in the photosynthetic zone, involves techniques which have developed in precision during the past 25 years. A method which is entirely useful either involves its being suitable for use on shipboard at the time the water samples are collected, or involves knowing how to store samples of water with certainty that no change takes place in the compound which is to be estimated. In either event the method of analysis needs to be quick, since very many analyses are almost always necessary in making any survey.

Methods of analysis of these and other microconstituents are the tools of chemical oceanography. It is by perfection of methods that future advances are likely.

Since many hundreds of such analyses are made in this laboratory every year, it is thought that an indication of some of the pitfalls experienced by us and the precautions taken may be of interest to others.

Inorganic orthophosphate. This is estimated by the molybdenum blue method. The estimate includes the equivalent of any arsenate which may be present, and it includes any acid soluble particulate phosphate.

The method is empirical.

In order to obtain precise estimates it is therefore necessary to adhere rigidly to a set of experimental conditions, which have been found satisfactory by experience, or to make the necessary corrections for changes in those experimental conditions which affect the quantity of blue formed or its rate of formation. In order to illustrate the effect of changing the reaction conditions and of various precautions, the procedure which has been in use for several years in this laboratory is described with the necessary detail.

Reagents used

(1) *Acid molybdate:* 6·5 g. ammonium molybdate is dissolved in 500 cm^3 distilled water plus 110 cm^3 concentrated sulphuric acid, and is stored in hard glass in the dark. It is ready for use after 48 hours; thereafter it undergoes little or no change.

(2) Stock solution of 40 g. clean crystals of stannous chloride dissolved in 50 cm^3 concentrated hydrochloric acid plus 50 cm^3 of distilled water. This very slowly oxidizes during long storage.

(3) *Stannous chloride,* 0·05N in 5 % HCl, ready for use during the following six hours: stock solution is diluted (about 1 in 50) with 5 % v/v hydrochloric acid, titrated with standard iodine and adjusted to 0·05N (1 cm^3 then contains 3 mg. Sn and decolorizes 2·5 cm^3 0·02N iodine).

(4) 0·0001 *molar* KH$_2$PO$_4$ *in* 0·28N *sulphuric acid:* made as required by diluting a 0·01 molar stock solution, in 0·28N sulphuric acid which has been saturated with chloroform, with 0·28N sulphuric acid.

Procedure (for clear offshore waters)

(i) Acid molybdate is added to the sample of sea water at room temperature in a hard glass flask in the proportion of 4·5 cm^3 to 100 cm^3.

(ii) The sample plus reagent is transferred to a 15 cm. cuvette or observation tube and its optical density S_0 measured, using an absorption meter with red light filters, or to a 10 cm. cuvette using a spectrophotometer at a wavelength near 700 mμ.

(iii) The sample plus reagent is returned to the flask and 0·2 cm^3 of 0·05N stannous chloride added (per 104 cm^3) while swirling the contents. The observation tube or cuvette is again filled and the optical density S_1 read when the colour has attained a maximum.

The time interval, after adding the reductant, during which 99·5 % of the maximum colour is present, varies with temperature at

11° C.	14–23 min.
14° C.	12–18 min.
16° C.	11–15 min.
23° C.	5–14 min.

(iv) The solution is at once returned to the flask and 0.500 ± 0.001 cm³ of 0.0001 molar phosphate added from a syringe pipette, followed by 0.2 cm³ of 0.05N stannous chloride. The increased optical density, S_2, is measured when it has reached its maximum value. This occurs in about three-quarters of the time needed for S_1.

The value of $S_2 - S_1$ is the increment in colour due to 15.5 mg. or 0.5 mg.-at. of phosphate-P per m³ at $t°$ C.

Alternatively, the increment can be determined by repeating operations (i)–(iii) with sea water to which phosphate has been added.

The increment in colour per unit addition of phosphate varies (a) with temperature, and increases 1.2% per $1°$ C. rise in temperature; (b) in waters of salinity less than $25\%_0$, in which it is greater; (c) with the quantity of stannous chloride added—an additional 0.03 cm³ of the 0.05N solution increases the increment by about 1%; (d) with the spectral composition of the band of red light used; (e) with any considerable quantity of particulate matter in suspension—thus bacteria or kaolin in sufficient quantity to make the water visibly cloudy reduced the increment by 4–5%; and (f) it is affected by additional acid or alkali. The colour developed is affected by change in acid and molybdate concentrations (from the customary 0.28N and 0.05%).

Ammonium molybdate present (%)	0·25	0·28	0·31 N-H_2SO
	Colour developed		
0·06	119	112	107
0·05	105	100	94
0·04	97	93	86

Under the experimental conditions, using Chance OR$_1$ or Ilford 608 red filters, change in optical density—absorbency or extinction—is directly proportional to phosphate concentration, provided no arsenate is present. Equivalent quantities of arsenate and of phosphate generate the same optical density.

The increment at a particular temperature is constant for different sea waters with salinity above $25\%_0$.

(v) In order to calculate the concentration of phosphate in the sample it is necessary to find the reagent blank. 4.5 cm³ acid molybdate is added to 100 cm³ distilled water. The observation tube or cuvette is very thoroughly washed to rid it of any trace of stannous chloride and the optical density, d_0, when filled with the solution is determined. The distilled water should not have been stored in soft glass, from which it has been found to dissolve small but significant quantities of phosphate.

Stannous chloride is then added as in operation (iii) and the optical density d_1 of the solution determined after the blue colour has developed. The colour developed and the interval during which 99·5 % of the colour is present, are both greater than with sea water. With distilled water

at   9°   45–145 min.
at 15°   15–45 min.
at 20°   10–30 min.

The reagent blank ($d_1 - d_0$ for $t°$ C.), being very small, the effect of salinity on colour formation is neglected. Then the concentration of phosphate-P in the water sample is

$$\frac{(S_1 - S_0) - (d_1 - d_0)}{(S_2 - S_1)} \times 15·5 \text{ mg. P per m}^3$$

or $$\frac{(S_1 - S_0) - (d_1 - d_0)}{(S_2 - S_1)} \times 0·5 \text{ mg.-at. P per m}^3$$

The reagents may be considered satisfactory if the value of $d_1 - d_0$ does not exceed the increment due to the addition of 1·0 mg. or 0·03 mg.-at. phosphate-P per m³ to sea water. High values have been found which were due to impure ammonium molybdate.

The difference between duplicate samples of raw water collected offshore very rarely exceed 1 mg. or 0·03 mg.-at. phosphate-P per m³ The differences between subsamples of a filtered sea water lay within ± 0·25 mg. P per m³

Notes

The stock solution of stannous chloride is diluted with 5 % HCl, because if diluted with water a faint turbidity is formed after addition to distilled water plus acid molybdate.

The addition of phosphate is made in $0·28\text{N-H}_2\text{SO}_4$ in order to avoid temporary localized acidities which could allow the formation of silico-molybdate.

The sample, plus acid molybdate, is kept well mixed by swirling while adding the $0 \cdot 2$ cm^3 of stannous chloride, in order to avoid irregular colour formation.

The value of S_0 is a measure of the turbidity of the water and so may vary from sample to sample, and is rarely negligible.

It is obvious that scrupulous cleanliness is necessary in collection of the samples, in washing stannous chloride from the flasks and keeping them free from dust.

Contamination with copper

The presence of this element depresses the formation of molybdenum blue. Samples should be kept out of contact with copper or brass from which variable quantities of copper ions may be discharged into the water.

The depression of optical density caused by the addition of a copper salt has been determined:

Copper (added as sulphate) (mg. per m^3)	0	20	50	100
Optical density expressed as percentage	100	97	93·5	87·5

Inorganic phosphate in turbid waters

Turbid inshore waters such as those of the southern North Sea contain considerable quantities of solid phosphate in suspension. Some of this dissolves when samples are made acid with acid molybdate; the longer they stand before adding the stannous chloride, the greater is the concentration of phosphate found.

Storage of samples for subsequent phosphate estimation

When a sample of sea water is stored, the phosphate concentration increases either from the start or after a short temporary decrease. Saturation with chloroform reduces these changes (p. 43); saturation with chloroform and storage in a refrigerator appears the most practical method of preserving samples for a few days.

It has been observed that storage for long periods in (some) soft glass bottles allowed considerable solution of phosphate from the glass.

Collier and Marvin (1953) have tested a method of storage for long periods. On collection the samples were filled into glass tubes, sealed with plastic caps and immersed in a mixture of solid carbon dioxide and alcohol or in ethylene glycol at $-20°$ C., then stored at subzero temperature until required for analysis.

Total and organic phosphorus.

This has been estimated as phosphate in sea water, and in water filtered free from plankton and detritus, by three methods:

A. After decomposition of organic phosphorus with sulphuric acid and hydrogen peroxide (Redfield, Smith and Ketchum, 1937).

B. After decomposition with perchloric acid (Hansen and Robinson, 1953).

C. After making acid to 0.28N with H_2SO_4 by hydrolysis at $135-140°$ C. for 5–6 hours (Harvey, 1948).

In methods A and B high and variable blank values were found, with considerable variation between duplicate samples. However, in subsequent analyses (A, B. H. Ketchum, private communication; B, F. A. J. Armstrong, in the press), where the decompositions were conducted in silica vessels, this was overcome, and in each series close agreement was found with values obtained by method C, which is simpler where many samples need to be analysed.

Arsenic does not interfere in these methods, being volatilized as chloride in method B, while arsenate is reduced to arsenite in methods A and C.

Method C, the determination of total phosphate after hydrolysis, will now be described.

Reagents used

(1) 50/50 *sulphuric acid:* equal volumes of concentrated acid and distilled water.

(2) *Sodium sulphite:* saturated solution.

(3) *Ammonium molybdate:* 6·6 g. dissolved in 400 cm.³ distilled water (free from silicate). Store in polythene bottle. The water needs to be freshly distilled and kept out of contact with soft glass or ground glass.

(4) 0.05 *stannous chloride, in* 5 % *v/v hydrochloric acid,* fresh as for inorganic phosphate.

(5) *Standard* 0.0001 *molar* KH_2PO_4, as for inorganic phosphate.

(6) *Thorium carbonate suspension:* add a 5 % solution of ammonium carbonate to 6 % solution of thorium nitrate until neutral to phenol red which is used as an external indicator.

Silica flasks

Short wide-necked transparent silica flasks (Vitreosil) of 100 cm³ capacity, cleaned with hot concentrated sulphuric acid, have been found satisfactory for the digestion. A similar set of flasks have been found unsatisfactory until they had been boiled in 10 % sodium hydroxide (J. H. Oliver, private communication). Phosphate dissolves from soft or borosilicate glass into hot acid.

Storage of samples prior to analysis

When samples of sea water were filled into silica flasks at the time of collection, more total phosphorus was found than when samples were first filled into glass bottles and, after storage, transferred to the silica flasks.

The difference is due to (i) growth of bacteria attached to the walls of the bottle, (ii) adsorption of dissolved organic phosphorus compounds on the glass. Solution of phosphate from glass—certainly from some, notably soft, glass—into the water may reduce this difference.

Mean concentration of total P in mg. P per m³

	Experiment A	Experiment B
Water filled into silica flasks at time of collection	14·0 (4)	15·5 (9)
After storage in glass bottles in darkness	10·5 (4)	14·0 (5)
After storage in baited bottles	14·7 (4)	16·0 (5)

(The numbers in parentheses are the number of analyses of subsamples of raw water from which the mean is derived.)

Samples have been stored for one to two weeks with very little change by means of the following trick. Bottles, which have been 'baited' previously with a few drops of chloroform and 1·0 cm³ thorium carbonate suspension, are charged at the time of collection with 67 cm³ of sea water. The chloroform stops bacterial action, and the fine particles of carbonate present a large surface for adsorption, acting as a collector which is subsequently dissolved.

Another method of storage, in use at the Woods Hole Oceano-
graphic Institute, is to charge bottles with about 250 cm³ of
water at the time of collection and add 1 cm³ of concentrated
hydrochloric acid. When required for analysis the inside surfaces
of the bottles are very thoroughly rubbed with a rubber-tipped
rod, before the contents are transferred to silica flasks. No
solution took place into the sea water from the glass of these
bottles (B. H. Ketchum, private communication).   ★

Procedure

The silica flasks are charged with 67 cm³ of sea water at the
time of collection. 1·00 cm³ of 50/50 sulphuric acid and 4 drops
of sodium sulphite (to reduce arsenate) are added before
autoclaving.

Alternatively, the contents of the baited bottles are acidified
with 1·00 cm³ of 50/50 sulphuric acid, which quickly dissolves
the thorium carbonate, and at once transferred to the silica
flasks. Four drops of sodium sulphite are added to each.

The flasks are covered with inverted beakers and autoclaved
for 5–6 hours at 30–40 lb. per in²

After cooling, the solution in each flask is made up to 68 cm³
with distilled water, transferred to hard glass flasks, and
2·00 cm³ of molybdate solution added with a hard glass pipette,
the contents being kept swirling during the addition.

After standing at least three minutes, phosphate may be
estimated by the procedure for inorganic phosphate from
step (ii) onwards.

The reagent blank (including that of the thorium carbonate)
may be estimated without autoclaving, although it is a useful check to autoclave one or two flasks for reagent blank
occasionally.

Notes

The increment in optical density due to unit addition of phosphate is about 4 % less in sea water which has been 'baited'
with thorium carbonate. Allowance for this is made auto-
matically in following the procedure for inorganic phosphate.

Experience has shown that the reagent blank $(d_1 - d_0)$ when
corrected for temperature, and the increment $(S_2 - S_1)$ per unit

addition of phosphate) after correction for temperature, remain constant from day to day using the same batches of reagents.

Differences in total phosphorus between duplicate analyses of raw waters, due for the most part to plankton organisms, do not often exceed 2 mg. (or 0·06 mg.-at.) P per m³, with occasional notably high values due for instance to the inclusion of a copepod. Analyses of nine subsamples of a filtered sea water varied between 6·5 and 7·2 mg., a spread of 0·7 mg. (0·02 mg.-at.) per m³

Estimation of the organic phosphorus is obtained by difference between total and inorganic phosphate, and contains also any particulate inorganic which did not dissolve in acid during the estimation of inorganic phosphate. If the water contained any significant quantity of arsenate, the equivalent of this appears in the analysis of inorganic phosphate, but not in the analysis of total phosphorus.

Phosphorus in biological material. A method for estimating phosphorus, within the range of 0 to 70 μg. P, in small quantities of phytoplankton or zooplankton collected at the bottom of a centrifuge tube, has proved satisfactory (Harvey, 1953).

★ **Ammonia.** Ammonium-N in sea water has been estimated directly with Nessler's reagent after precipitating or after chelating calcium and magnesium (Wattenberg, 1937). Another direct method makes use of the reaction of ammonia with hypobromite and gives better agreement between duplicate samples (Buljan, 1951).

In the method due to Krogh (1934) the sea water sample at pH 10–11 is distilled under reduced pressure and the ammonia estimated by reaction with hypobromite, the excess of which is then titrated with naphthyl red. The end-point is rather indistinct and duplicates do not always agree well.

Another method has been developed by J. P. Riley (1953). The ammonia in the distillate from sea water brought to pH 9·15 with metaborate is estimated by the colour produced with hypochlorite and phenate when catalysed by manganese. Under the experimental conditions used, Beer's law is obeyed with concentrations up to 2 mg. ammonia-N per litre in the distillate. No ammonia is produced from amino acids or urea. Replicate estimations for a particular sea water were in good agreement, with a standard deviation of less than 1 mg. ammonium-N per m.[3]

When sea water is stored considerable changes take place in its ammonia content (Redfield and Keys, 1938).

Nitrite. Estimation of nitrite presents no difficulty, using a Griess reaction. The details of reaction times, etc., using various techniques have been compared by Barnes and Folkard (1951). (See also Benschneider and Robinson, 1952.)

Changes in nitrite and nitrate content on storage of samples collected close to the sea bottom have been observed. (See also Barkova et al. 1936.)

Nitrate.

Approximate estimation with reduced strychnine (strychnidine)

⁻ When strychnine reduced with zinc and hydrochloric acid, or strychnidine which is readily obtained by electrolytic reduction, is dissolved in concentrated sulphuric acid and added to a solution containing chloride, and nitrate or nitrite at great dilution, a rose red colour is produced. This reaction allows an approximate estimation of nitrate and nitrite in sea waters (Harvey, 1928; Cooper, 1933; Deacon, 1933; and others).

An examination of the reactions involved (Armstrong, unpublished) has led to the conclusion that the oxidation of strychnidine in sea water containing nitrate is a chain reaction in which dissolved oxygen takes part. More oxidized strychnidine (the red compound) is produced than is stoichiometrically equivalent to the nitrate originally present, as was observed by Zwicker and Robinson (1944). If the reagent and the sea water are deoxygenated by bubbling nitrogen through them, on mixing only a faint pink colour results.

On mixing reagent and a sea water with varying additions of nitrate, the colour formed is directly proportional to the nitrate. Equivalent quantities of nitrite and of nitrate give the same intensity of red colour when the reaction takes place in stoppered bottles, precluding loss of oxides of nitrogen.

The presence of appreciable amounts of plankton organisms reduces the colour formation. The intensity of colour formed is very sensitive to conditions during the mixing and varies with age of the reagent.

From the numerous concordant results, it is apparent that the estimates obtained by this method in clear waters from the open sea have provided a good general picture of the considerable variations in nitrate content which occur regularly during the passage of the seasons, with depth and with geographical position in the oceans. This method of estimation has played a useful part and now needs to be replaced by one which is more precise and certain.

Reduction to nitrite and estimation with Griess reagent

Føyn (1951) heats 100 cm³ of sea water with 40 g. granulated zinc under defined conditions, decants and estimates nitrite. About 10 % of the nitrate is converted to nitrite. The intensity of colour produced in six subsamples indicated concentrations varying by ± 4·5 mg. nitrate-N per m³

Polarographic determination

The technique described by Chow and Robinson (1953) gave values which agreed within ± 2·8 mg. nitrate-N per m³ The method permits storage of the samples with chloroform.

If a method could be devised of estimating the total available nitrogen, present in a water sample as ammonia nitrite and nitrate (also amide-N ?), quickly and accurately, in one operation, this would be valuable. Each ★ is equally available for plant production.

Organic nitrogen in biological material. A method of estimating Kjeldahl nitrogen within the range of 3–30 μg. without distillation (Harvey, 1951) has proved satisfactory for small quantities of phytoplankton collected in a centrifuge tube.

Silicate. A method in which silico-molybdenum blue is measured with an absorption meter, covering the range between 0·1 and 50 mg.-at. Si per m³ has been used extensively (Armstrong, 1951), and found valuable for distinguishing water masses and deducing their provenance. The colour development on which the method is based is reasonably constant. In this laboratory, using the same absorption meter, calibration experiments over a period of 3 years have shown $E_{10\,cm.} = 0.181 \pm 0.005$, for an addition of 2·00 mg.-at. silicate-Si per m? (1 mg.-at. Si = 28 mg. Si = 60 mg. SiO_2.) Various methods based on the measurement of yellow silicomolybdate have been in use. Two species of silicomolybdic acid may be formed, depending upon experimental conditions (Strickland, 1952). When a reproducible mixture is formed under standardized conditions, extinction and concentration are not linearly proportional unless monochromatic light is used.

A method in which yellow silicomolybdate is measured with a spectrophotometer at 430 mμ has been described by Chow and Robinson (1953). Other wavelengths for the measurement have been proposed. The maximum extinction of the silicomolybdate is in the ultraviolet, but overlaps that of the excess molybdate. In order to obtain reasonably small blank extinctions it is necessary to choose a wavelength on the skirt of the absorption curve. Some sensitivity is necessarily sacrificed and for good reproducibility the wavelength setting must be exact. Most of these difficulties are overcome by reducing to molybdenum blue.

References to analytical methods, such as determination of salinity, titration alkalinity, oxygen content, the carbon dioxide equilibrium system, iron and manganese are given in the text. The bibliography dealing with the occurrence of other constituents provides information of how they have been estimated.

A valuable manual of techniques for the analyses of several constituents of seawaters has been published by Strickland (*Standard Methods of Sea Water Analysis*, Fish Res. Board Canada, Pacific Oceanography Group, Nanaimo B.C., also Manuscript Report series No. 19, Fish Res. Board Canada, 1957).

200

BIBLIOGRAPHY

ADENEY,W.E.(1928). *The Dilution Method of Sewage Disposal.* Cambridge.

ALEXANDER, W. B., SOUTHGATE, B. A. and BASSINDALE, R. (1935). Survey of the River Tees, Part II. *Water Poll. Res., Tech. Paper No. 5.* London, H.M.S.O.

ALLEN, E. J. (1914). On the culture of the plankton diatom *Thalassiosira gravida* in artificial sea-water. *J. Mar. Biol. Ass. U.K.* **10**, 417.

ALLEN, E. J. and NELSON, E. W. (1910). On the artificial culture of marine plankton organisms. *J. Mar. Biol. Ass. U.K.* **8**, 421.

ARMSTRONG, F. A. J. (1951). The determination of silicate in sea water. *J. Mar. Biol. Ass. U.K.* **30**, 149.

ARMSTRONG, F. A. J. (1954). Phosphorus and silicon in sea water off Plymouth during the years 1950 to 1953. *J. Mar. Biol. Ass. U.K.* **33**, 382.

ARMSTRONG, F. A. J. and HARVEY, H. W. (1950). The cycle of phosphorus in the waters of the English Channel. *J. Mar. Biol. Ass. U.K.* **29**, 145.

ARRHENIUS, G. (1952). Sediment cores from the East Pacific. *Rep. Swedish Deep-Sea Exped.* **5**, parts 1 and 2.

ATKINS, W. R. G. (1922). Respirable organic matter of sea water. *J. Mar. Biol. Ass. U.K.* **12**, 772.

ATKINS, W. R. G. (1923-30). The silica content of some natural waters and of culture media. *J. Mar. Biol. Ass. U.K.* **13**, 151. See also **14**, 89; **15**, 191; **16**, 821.

ATKINS, W. R. G. (1932). The copper content of sea-water. *J. Mar. Biol. Ass. U.K.* **18**, 193.

ATKINS, W. R. G. (1936). The estimation of zinc in sea water. *J. Mar. Biol. Ass. U.K.* **20**, 625.

ATKINS, W. R. G. (1938). Photoelectric measurements of the seasonal variations in daylight. *Proc. Roy. Soc. A*, **165**, 453.

ATKINS, W. R. G. (1953). The seasonal variation in the copper content of sea water. *J. Mar. Biol. Ass. U.K.* **31**, 493.

ATKINS, W. R. G. and WILSON, E. G. (1926). The colorimetric estimation of minute amounts of compounds of silicon, phosphorus and arsenic. *Biochem. J.* **20**, 1223.

ATKINS, W. R. G. and WILSON, E. G. (1927). The phosphorus and arsenic compounds of sea-water. *J. Mar. Biol. Ass. U.K.* **14**, 609.

BACHRACH, E. and LUCCIARDI, N. (1932). Influence de la concentration en ions hydrogène (pH) sur la multiplication de quelques diatomées marines. *Rev. algol.* **6**, 251.

BALDWIN, E. (1949). *An Introduction to Comparative Biochemistry.* Cambridge University Press.

BARDET, J., TCHAKIRIAN, A. and LAGRANGE, R. (1938). Recherche spectrographique des éléments existant a l'état de traces dans l'eau de mer. *C.R. Acad. Sci., Paris*, **206**, 450.

BARKER, H. A. (1935a). Photosynthesis in diatoms. *Arch. Mikrobiol.* **6**, 141.

BARKER, H. A. (1935b). The culture and physiology of marine dino-flagellates. *Arch. Mikrobiol.* **6**, 157.

BARKOVA, E., BOVSOOK, V., VERJBINSKAYA, N., KREPS, E. and LUK-YANOWA, V. (1936). Enzymes in sea water. *Arch. Sci. Biol. U.S.S.R.* **43**, 362.

BARNES, H. and FOLKARD, A. R. (1951). The determination of nitrites. *Analyst*, **76**, 599.

BATHER, J. M. and RILEY, J. P. (1953). The precise and routine poten-tiometric determination of the chlorinity of sea water. *J. Cons. int. Explor. Mer*, **18**, 277.

BEHARRELL, J. (1942). Seaweed as a food for livestock. *Nature, Lond.*, **149**, 306.

BENECKE, W. (1933). Bakteriologie des Meeres. *Abderhaldens Handb. biol. Meth.* Abt. IX, Teil 5, Meerwasserbiologie, **1**, 717.

BENDSCHNEIDER, K. and ROBINSON, R. J. (1952). A new spectro-photometric method for the determination of nitrite in sea water. *J. Mar. Res.* **11**, 87.

BENNETT, M. and RIDEAL, E. (1954). Membrane behaviour in *Nitella*. *Proc. Roy. Soc. B*, **142**, 483.

BERNARD, F. (1939). Étude sur les variations de fertilité des eaux méditerranéennes. *J. Cons. int. Explor. Mer*, **14**, 228.

BERTEL, R. (1912). Sur la distribution quantitative des bactéries planctoniques des côtes de Monaco. *Bull. Inst. océanogr. Monaco*, no. 224.

BIGELOW, H. B., LILLICK, L. and SEARS, M. (1940). Phytoplankton and planktonic protozoa of the offshore waters of the Gulf of Maine. Pt. I. *Trans. Amer. Phil. Soc.* **31**, 149.

BIRGE, E. A. and JUDAY, C. (1934). Particulate and dissolved organic matter in inland lakes. *Ecol. Monogr.* **4**, 440.

BLACK, W. A. P. and MITCHELL, R. L. (1952). Trace elements in the common brown algae and in sea water. *J. Mar. Biol. Ass. U.K.* **30**, 575.

BOHR, C. (1899). Definition und Methode zur Bestimmung der Invasions-und Evasionskoefficienten bei der Auflösung von Gasen in Flüssig-keiten. *Ann. Phys. Chem.* **68**, 500.

BOURY, M. (1938). Le plomb dans le milieu marin. *Rev. Trav. Off. Pêches marit.* **11**, 157.

BRAADLIE, O. (1930). Innholdet av ammoniakk og nitratvelstoff i nedbøren ved Trondhjem. *K. norske vidensk. Selsk. Forh.* **3**, no. 20.

BRAARUD, T. (1945). Experimental studies on marine plankton diatoms. *Avh. norske VidenskAkad*, 1944, no. 10, 16 pp.

BRAARUD, T. and FØYN, B. (1931). Beiträge zur Kenntnis des Stoff-wechsels im Meere. *Avh. norske VidenskAkad.* 1930, no. 14, 24 pp.

BRAARUD, T., GAARDER, K. R. and GRØNTVED, J. (1953). The phyto-plankton of the North Sea and adjacent waters in May, 1948. *Rapp. Cons. Explor. Mer*, **133**, 1.

BRAARUD, T. and KLEM, A. (1931). Hydrographical and chemical investigations in the coastal waters off Møre. *Hvalråd. Skr.* no. 1, 88 pp.

VON BRAND, T. (1938). Quantitative determination of the nitrogen in the particulate matter of the sea. *J. Cons. int. Explor. Mer*, **13**, 187.

VON BRAND, T. and RAKESTRAW, N. W. (1941). The determination of dissolved organic nitrogen in sea water. *J. Mar. Res.* **4**, 76.

VON BRAND, T. and RAKESTRAW, N. W. (1937–42). Decomposition and regeneration of nitrogenous organic matter in sea water. *Biol. Bull., Woods Hole*, **72**, 165; **77**, 285; **79**, 231; **81**, 63; **83**, 273.

BRUNEAU, L., JERLOV, N. G. and KOCZY, F. F. (1953). 'Physical and Chemical Methods.' *Rep. Swedish Deep-Sea Exped.* 1947–8, **3**, 99.

BUCH, K. (1929a). Die Verwendung von Stickstoff- und Sauerstoff-analysen in der Meeresforschung. *J. Cons. int. Explor. Mer*, **4**, 162.

BUCH, K. (1929b). Ueber den Einfluss der Temperatur auf die pH-Bestimmung des Meerwassers. *HavsforsknInst. Skr., Helsingf.*, no. 61.

BUCH, K. (1933a). Der Borsäuregehalt des Meerwassers und seine Bedeutung bei der Berechnung des Kohlensäuresystems in Meer-wasser. *Rapp. Cons. Explor. Mer*, **85** pt. III, 71.

BUCH, K. (1933b). On boric acid in the sea and its influence on the carbonic acid equilibrium. *J. Cons. int. Explor. Mer*, **8**, 309.

BUCH, K. (1938). New determination of the second dissociation constant of carbonic acid in sea water. *Acta Acad. åbo*, Math. et Phys., **11**, no. 5.

BUCH, K. (1939a). Beobachtungen über das Kohlensäuregleichgewicht und über den Kohlensäureaustausch zwischen Atmosphäre und Meer im Nordatlantischen Ozean. *Acta Acad. åbo*, Math. et Phys., **11**, no. 9.

BUCH, K. (1939b). Kohlensäure im Atmosphäre und Meer an der Grenze zum Arktikum. *Acta Acad. åbo*, Math. et Phys., **11**, no. 12.

BUCH, K. (1951). Das Kohlensäure gleichgewichtssystem im Meer-wasser. *HavsforsknInst. Skr., Helsingf.*, no. 151.

BUCH, K. (1952). The cycle of nutrient salts and marine production. *Rapp. Cons. Explor. Mer*, **132**, 36.

BUCH, K. and GRIPENBERG, S. (1932). Ueber den Einfluss des Wasser-druckes auf pH und das Kohlensäuregleichgewicht in grösseren Meerestiefen. *J. Cons. int. Explor. Mer*, **7**, 233.

BUCH, K., HARVEY, H. W., WATTENBERG, H. and GRIPENBERG, S. (1932). Ueber das Kohlensäuresystem in Meerwasser. *Rapp. Cons. Explor. Mer*, **79**, 1.

BUCH, K. and NYNÄS, O. (1939). Studien über neuere pH Methodik usw. *Acta Acad. åbo*, Math. et Phys., **12**, no. 3.

BUCH, K. and URSIN, M. (1948). Zur Methodik der Bestimmung von Phosphor im Meerwasser. *Havsforskn. Inst. Skr. Helsingf.* **140**, 1.

BULJAN, M. (1951). Note on a method for determination of ammonia in sea water. *J. Mar. Biol. Ass. U.K.* **30**, 277.

CAREY, C. L. (1938). The occurrence and distribution of nitrifying bacteria in the sea. *J. Mar. Res.* **1**, 291.

CAUSEY, G. and HARRIS, E. J. (1951). The uptake and loss of phosphate by frog muscle. *Biochem. J.* **49**, 176.

CHOLODNY, N. (1929). Zur Methodik der quantitativen Erforschung des bakteriellen Planktons. *Zbl. Bakt. Abt.* 11, 77, 179.

CHOW, D. T. W. and ROBINSON, R. J. (1953). Forms of silicate available for colorimetric determination. *Analyt. Chem.* **25**, 646.

CHOW, D. T. W. and ROBINSON, R. J. (1953). Polarographic determination of nitrate in sea water. *J. Mar. Res.* **12**, 1.

CHOW, T. J. and THOMPSON, T. G. (1952). The determination and distribution of copper in sea water. Pt. I. *J. Mar. Res.* **11**, 124.

CHU, S. P. (1949). Experimental studies on the environmental factors influencing the growth of phytoplankton. *Science and Technology in China*, **2**, no. 3, p. 38.

CLOWES, A. J. (1938). Phosphate and silicate in the Southern Ocean. '*Discovery*' *Rep.* **19**, 1.

COLLIER, A. and MARVIN, K. T. (1953). Stabilization of the phosphate ratio of sea water by freezing. *Fish. Bull. U.S.* **79**.

COLLIER, A., RAY, S. M., MAGNITZKY, A. W. and BELL, I. O. (1953). Effect of dissolved organic substances on oysters. *Fish. Bull. U.S.* **84**.

COMERE, J. (1909). Action of arsenates on the growth of algae. *Bull. Soc. Bot. Fr.* **56**, 147.

COOPER, L. H. N. (1933). Chemical constituents of biological importance in the English Channel. Pt. I. Phosphate, silicate, nitrate, nitrite, ammonia. *J. Mar. Biol. Ass. U.K.* **18**, 677.

COOPER, L. H. N. (1935a). Iron in the sea and in marine plankton. *Proc. Roy. Soc.* B, **118**, 419.

COOPER, L. H. N. (1935b). Liberation of phosphate in sea water by the breakdown of plankton organisms. *J. Mar. Biol. Ass. U.K.* **20**, 197.

COOPER, L. H. N. (1937a). The nitrogen cycle in the sea. *J. Mar. Biol. Ass. U.K.* **22**, 183.

COOPER, L. H. N. (1937b). Oxidation-reduction potential in sea water. *J. Mar. Biol. Ass. U.K.* **22**, 167.

COOPER, L. H. N. (1937c). Some conditions governing the solubility of iron. *Proc. Roy. Soc.* B, **124**, 299.

COOPER, L. H. N. (1937d). On the ratio of nitrogen to phosphorus in the sea. *J. Mar. Biol. Ass. U.K.* **22**, 177.

COOPER, L. H. N. (1938a). Redefinition of the anomaly of the nitrate-phosphate ratio. *J. Mar. Biol. Ass. U.K.* **23**, 179.

COOPER, L. H. N. (1938b). Salt error in determinations of phosphate in sea water. *J. Mar. Biol. Ass. U.K.* **23**, 171.

COOPER, L. H. N. (1938c). Phosphate in the English Channel. *J. Mar. Biol. Ass. U.K.* **23**, 181.

COOPER, L. H. N. (1948a). The distribution of iron in the waters of the western English Channel. *J. Mar. Biol. Ass. U.K.* **27**, 279.

COOPER, L. H. N. (1948b). Some chemical considerations on the distribution of iron in the sea. *J. Mar. Biol. Ass. U.K.* **27**, 314.

COOPER, L. H. N. (1951). Chemical properties of the sea water in the neighbourhood of the Labadie Bank. *J. Mar. Biol. Ass. U.K.* **30**, 21.

COOPER, L. H. N. (1952). Factors affecting the distribution of silicate in the North Atlantic. *J. Mar. Biol. Ass. U.K.* **30**, 511.

CORLETT, J. (1953). Net phytoplankton at ocean weather stations I and J. *J. Cons. int. Explor. Mer*, **19**, 178.

DAKIN, W. J. and COLEFAX, A. N. (1935). Observations on the seasonal changes in temperature, salinity, phosphates, and nitrate nitrogen and oxygen of the ocean waters off New South Wales. *Proc. Linn. Soc. N.S.W.* **60**, 303.

DANIELLI, J. F. (1944). The biological action of ions and the concentration of ions at surfaces. *J. Exp. Biol.* **20**, 167.

DATZKO, W. (1951). The vertical distribution of organic matter in the Black Sea. *C.R. Acad. Sci. U.R.S.S.* **77**, 1059.

DEACON, G. E. R. (1933). A general account of the hydrology of the South Atlantic Ocean. '*Discovery*' *Rep.* **7**, 171.

DEACON, G. E. R. (1940). Carbon dioxide in the Arctic and Antarctic Seas. *Nature, Lond.*, **145**, 250.

DESGREZ, A. and MEUNIER, J. (1926). Recherche et dosage du strontium dans l'eau de mer. *C.R. Acad. Sci.*, *Paris*, **183**, 689.

DEVAPUTRA, D., THOMPSON, T. G. and UTTERBACK, C. L. (1932). The radioactivity of sea water. *J. Cons. int. Explor. Mer*, **7**, 358.

DIETZ, R., EMERY, K. and SHEPARD, F. (1942). Phosphorite deposits on the sea floor off Southern California. *Bull. Geol. Soc. Amer.* **53**, 815.

DOMOGALLA, B. P., JUDAY, C. and PETERSON, W. H. (1925). The forms of nitrogen found in certain lake waters. *J. Biol. Chem.* **63**, 269.

DUTTON, H., MANNING, W. and DUGGAR, B. (1943). Chlorophyll fluorescence and energy transfer in the diatom *Nitzschia closterium*. *J. Phys. Chem.* **47**, 308.

EKMAN, V. W. (1910). Tables for sea-water under pressure. *Publ. Circ. Cons. Explor. Mer*, no. 49. International Council for the Exploration of the Sea.

ERCEGOVIC, A. (1934). Température, salinité, oxygène et phosphates dans les eaux cotières de l'Adriatique. *Acta adriat.* no. 5.

VON ENGELHARDT, W. (1936). Die Geochemis des Bariums. *Chem. d. Erde*, **10**, 187.

ERNST, T. and HOERMANN, H. (1936). Bestimmung von Vanadium, Nickel und Molybdän im Meerwasser. *Nachr. Ges. Wiss. Göttingen*, Math. Phys., **1**, 205.

EVANS, R., KIP, A. and MOBERG, E. (1938). The radium and radon content of Pacific Ocean water, life and sediments. *Amer. J. Sci.* **36**, 241.

FLEMING, R. H. (1940). Composition of plankton. *Proc. 6th Pacific Sci. Congr.* **3**, 535.

FOGG, G. E. (1953). *The Metabolism of Algae*. London, New York: Methuen.

FOX, C. J. J. (1907). On the coefficients of absorption of atmospheric gases in sea water. *Publ. Circ. Cons. Explor. Mer*, no. 41.

FOX, H. M. and WINGFIELD, C. A. (1938). A portable apparatus for the determination of oxygen dissolved in a small volume of water. *J. Exp. Biol.* **15**, 437.

Føyn, E. (1951). Nitrogen determinations in sea-water. *Fiskeridir. Skr. Havundersøk.* **9**, no. 14.

Føyn, E., Karlik, B., Pettersson, H. and Rona, E. (1939). Radioactivity in sea water. *Medd. oceanog. Inst. Göteborg*, **6**, no. 12.

Friedman, I. (1953). Deuterium content of natural waters. *Geochim. et cosmoch. Acta*, **4**, 89.

Gardiner, A. C. (1937). Phosphate production by planktonic animals. *J. Cons. int. Explor. Mer*, **12**, 144.

Gee, H., Greenberg, D. M. and Moberg, E. G. (1932). Calcium equilibrium in sea water. II. Sealed bottles shaken at constant temperature. *Bull. Scripps Instn Oceanogr.* **3**, 158.

Gillbricht, M. (1952). Untersuchungen zur Produktions-biologie des Planktons in der Kieler Bucht. *Kieler Meeresforsch.* **8**, 173.

Goldberg, E. D. (1952). Iron assimilation by marine diatoms. *Biol. Bull., Woods Hole*, **102**, 243.

Goldberg, E. D., McBlair, W. and Taylor, K. M. (1951). The uptake of vanadium by tunicates. *Biol. Bull., Woods Hole*, **101**, 84.

Goldberg, E. D., Walker, T. J. and Whisenand, A. (1951). Phosphate utilization by diatoms. *Biol. Bull., Woods Hole*, **101**, 274.

Goldschmidt, V. M. (1934). Drei Vorträge über Geochemie. *Geol. Fören. Stockh. Förh.* **56**, 385.

Goldschmidt, V. M. (1937). The principles and distribution of chemical elements in minerals and rocks. *J. Chem. Soc.* p. 655.

Gorgy, S., Rakestraw, N. W. and Fox, D. L. (1948). Arsenic in the sea. *J. Mar. Res.* **7**, 22.

Greenberg, G., Moberg, E. and Allen, E. (1932). The determination of CO_2 and titratable base in sea water. *Industr. Engng Chem.* (Anal. ed.), **4**, 309.

Griel, J. V. and Robinson, R. J. (1952). Titanium in sea water. *J. Mar. Res.* **11**, 173.

Gripenberg, S. (1937a). The calcium content of Baltic Water. *J. Cons. int. Explor. Mer*, **12**, 293.

Gripenberg, S. (1937b). On the determination of excess base in sea water. *Comm.* 10B, *Vth Hydrol. Conf. Helsingfors*, 1936.

Gross, F. and Zeuthen, E. (1948). The buoyancy of plankton diatoms: a problem of cell physiology. *Proc. Roy. Soc.* B. **135**, 382.

Guelin, A. (1954). La contamination des poissons et le problème des eaux polluées. *Ann. Inst. Pasteur*, **86**, 303.

Haber, F. (1928). Das Gold im Meer. *Z. Ges. Erdk. Berl.* Erg. H. 3. See also Jaenicke, J. (1935). Habers Untersuchungen über das Gold im Meere. *Naturwissenschaften*, **23**, 57.

Haendler, H. M. and Thompson, T. G. (1939). Determination and occurrence of aluminium in sea water. *J. Mar. Res.* **2**, 12.

Hamilton, L., Hutner, G. and Provosoli, L. (1952). The use of Protozoa in analysis. *Analyst*, **77**, 618.

Hansen, A. L. and Robinson, R. J. (1953). The determination of organic phosphorus in sea water with perchloric acid oxidation. *J. Mar. Res.* **12**, 31.

HARDING, M. W. and MOBERG, E. G. (1934). Determination and quantity of boron in sea water. *Proc. Pacific Sci. Congr.* 1933, Canada, **3**, 2093–5.

HART, T. J. (1934). Phytoplankton of the south-west Atlantic. '*Discovery*' *Rep.* **8**, 185.

HART, T. J. (1942). Phytoplankton periodicity in Antarctic surface waters. '*Discovery*' *Rep.* **21**, 261.

HARVEY, H. W. (1925). Oxidation in sea water. *J. Mar. Biol. Ass. U.K.* **13**, 953.

HARVEY, H. W. (1928). Nitrate in the sea. *J. Mar. Biol. Ass. U.K.* **15**, 183.

HARVEY, H. W. (1933). On the rate of diatom growth. *J. Mar. Biol. Ass. U.K.* **19**, 253.

HARVEY, H. W. (1937a). The supply of iron to diatoms. *J. Mar. Biol. Ass. U.K.* **22**, 205.

HARVEY, H. W. (1937b). Note on colloidal ferric hydroxide in sea water. *J. Mar. Biol. Ass. U.K.* **22**, 221.

HARVEY, H. W. (1939). Substances controlling the growth of a diatom. *J. Mar. Biol. Ass. U.K.* **23**, 499.

HARVEY, H. W. (1940). Nitrogen and phosphorus required for the growth of phytoplankton. *J. Mar. Biol. Ass. U.K.* **24**, 115.

HARVEY, H. W. (1941). On changes taking place in sea water during storage. *J. Mar. Biol. Ass. U.K.* **25**, 225.

HARVEY, H. W. (1947). Manganese and the growth of phytoplankton. *J. Mar. Biol. Ass. U.K.* **26**, 562.

HARVEY, H. W. (1948). The estimation of phosphate and of total phosphorus in sea waters. *J. Mar. Biol. Ass. U.K.* **27**, 337.

HARVEY, H. W. (1949). On manganese in sea and fresh waters. *J. Mar. Biol. Ass. U.K.* **28**, 155.

HARVEY, H. W. (1950). On the production of living matter in the sea off Plymouth. *J. Mar. Biol. Ass. U.K.* **29**, 97.

HARVEY, H. W. (1951). Micro-determination of nitrogen in organic matter without distillation. *Analyst*, **76**, 657.

HARVEY, H. W. (1953a). Note on the absorption of organic phosphorus compounds by *Nitzschia closterium* in the dark. *J. Mar. Biol. Ass. U.K.* **31**, 475.

HARVEY, H. W. (1953b). Synthesis of organic nitrogen and chlorophyll by *Nitzschia closterium*. *J. Mar. Biol. Ass. U.K.* **31**, 477.

HARVEY, H. W. (1953c). Micro-determination of phosphorus in biological material. *Analyst*, **78**, 110.

HARVEY, H. W., COOPER, L. H. N., LEBOUR, M. V. and RUSSELL, F. S. (1935). Plankton production and its control. *J. Mar. Biol. Ass. U.K.* **20**, 407.

HAXO, F. T. and BLINKS, L. R. (1950). Photosynthetic action spectra of marine algae. *J. Gen. Physiol.* **33**, 389.

HAZEL, F. and AYERS, G. H. (1931). Migration studies with ferric oxide sols. I. Positive sols. *J. Phys. Chem.* **35**, 2930, 3148.

HENKEL, R. (1952). Ernährungsphysiologische Untersuchungen an Meeresalgen. *Kieler Meeresforsch.* **8**, 192.

HENTSCHEL, E. and WATTENBERG, H. (1930). Plankton und Phosphat in der Oberflächenschicht des Südatlantischen Ozeans. *Ann. Hydrogr.*, *Berl.*, **58**, 273.

HERNEGGER, F. and KARLIK, B. (1935). Uranium in sea-water. *Göteborgs VetenskSamh. Handl.*, Femte Foljden, **4**, no. 12.

HEUKELEKIAN, H. and HELLER, A. (1940). Relation between food concentration and surface for bacterial growth. *J. Bact.* **40**, 547.

HOCK, C. W. (1941). Marine chitin-decomposing bacteria. *J. Mar. Res.* **4**, 99.

HOTCHKISS, M. and WAKSMAN, S. A. (1936). Correlative studies of microscopic and plate methods for evaluating the bacterial population of the sea. *J. Bact.* **32**, 423.

HUTCHINSON, G. E. (1943). Thiamine in lake waters, etc. *Arch. Biochem.* **2**, 143.

HUTCHINSON, G. E. and SETLOW, J. K. (1946). Limnological studies in Connecticut. VIII. The Niacin cycle in a small inland lake. *Ecology*, **27**, 13.

HUTNER, S. H., PROVASOLI, L., SCHATZ, A. and HASKINS, C. P. (1950). Some approaches to the study of the role of metals in the metabolism of micro-organisms. *Proc. Amer. Phil. Soc.* **94**, 152.

IGELSRUD, I., THOMPSON, T. G. and ZWICKER, B. (1938). The boron content of sea water and marine organisms. *Amer. J. Sci.* **35**, 47.

ISHIBASHI, M. *et al.* (1951). The determination of arsenic in sea water. *Chem. Abstr.* **45**, 10443.

JACOBSEN, J. P. (1912). The amount of oxygen in the water of the Mediterranean. *Rep. Danish Oceanog. Exped.*, 1908–10, **1**, 209. Copenhagen.

JACOBSEN, J. P. and KNUDSEN, M. (1921). Dosage d'oxygène dans l'eau de mer par la méthode de Winkler. *Bull. Inst. océanogr. Monaco*, no. 390.

JACOBSEN, J. P. and KNUDSEN, M. (1940). Urnormal or primary standard sea-water. *Publ. sci. Ass. Océanogr. phys.* no. 7.

JENKIN, P. M. (1937). Oxygen production by the diatom *Goscinodiscus excentricus* in relation to submarine illumination in the English Channel. *J. Mar. Biol. Ass. U.K.* **22**, 301.

JERLOV, N. G. (1953). Particle distribution in the ocean. *Rep. Swedish Deep-Sea Exped.* 1947–8, **3**, 71.

JOHNSON, F. H. (1936). The oxygen uptake of marine bacteria. *J. Bact.* **31**, 547.

JONES, A. J. (1922). Arsenic content of some marine algae. *J. Pharmacol.* **109**, 86.

KALLE, K. (1937). Nährstoff-Untersuchungen als hydrographisches Hilfsmittel zur Unterscheidung von Wasserkörpern. *Ann. Hydrogr.*, *Berlin*, **65**, 1.

KALLE, K. (1949). Fluoreszenz und Gelbstoff in Bottnischen und Finnischen Meerbusen. *Dtsch. hydrogr. Z.* **2**, 117.

KALLE, K. and WATTENBERG, H. (1938). Über Kupfergehalt des Ozeanwassers. *Naturwissenschaften*, **26**, 630.

KETCHUM, B. H. (1939a). The absorption of phosphate and nitrate by illuminated cultures of *Nitzschia closterium*. *Amer. J. Bot.* **26**, 399.

KETCHUM, B. H. (1939b). The development and restoration of deficiencies in the phosphorus and nitrogen composition of unicellular plants. *J. Cell. Comp. Physiol.* **13**, 373.

KETCHUM, B. H., AYRES, J. C. and VACCARO, R. F. (1952). Processes contributing to the decrease of coliform bacteria in a tidal estuary. *Ecology*, **33**, 247.

KETCHUM, B. H., CAREY, C. L. and BRIGGS, M. (1949). Viability and dispersal of coliform bacteria in the sea. In *Limnological Aspects of Water Supply* (p. 64). Ed. by F. R. Moulton and F. Hitzel. Publ. Amer. Ass. Adv. Sci.

KETCHUM, B. H. and REDFIELD, A. C. (1949). Some physical and chemical characteristics of algae growth in mass culture. *J. Cell. Comp. Physiol.* **33**, 281.

KEYS, A. B. (1931). The determination of chlorides with the highest accuracy. *J. Chem. Soc.* p. 2440.

KEYS, A., CHRISTENSEN, E. H. and KROGH, A. (1935). Organic metabolism of sea-water. *J. Mar. Biol. Ass. U.K.* **20**, 181.

KIRK, P. L. and MOBERG, E. G. (1933). Microdetermination of calcium in sea water. *Industr. Engng Chem.* (Anal. ed.), **5**, 95.

KNOWLES, G. and LOWDEN, G. (1953). Methods for detecting the endpoint in the titration of iodine with thiosulphate. *Analyst*, **78**, 159.

KNUDSEN, M. (1901). *Hydrographical Tables.* Copenhagen.

KNUDSEN, M. (1902). Berichte über die Konstantenbestimmungen zur Aufstellung der hydrographischen Tabellen. *K. danske vidensk. Selsk.*, Math., 6 Raekke, **12**, 1.

KNUDSEN, M. (1903). Gefrierpunkttabelle für Meerwasser. *Publ. Circ. Cons. Explor. Mer*, no. 5.

KOCZY, F. F. (1949). Thorium in sea water and marine sediments. *Geol. Fören. Stockh. Förh.* **71**, 237.

KREPS, E. (1934). Organic catalysts or enzymes in sea water. *James Johnstone Memorial Volume*, pp. 193. Liverpool.

KROGH, A. (1910). Some experiments on the invasion of oxygen and carbonic oxide into water. The mechanism of gas exchange. V. *Skand. Arch. Physiol.* **23**, 224.

KROGH, A. (1934a). Conditions of life at great depths in the ocean. *Ecol. Monogr.* **4**, 430.

KROGH, A. (1934b). A method for the determination of ammonia in water and air. *Biol. Bull., Woods Hole*, **67**, 126.

KROGH, A. and KEYS, A. (1934). Methods for the determination of dissolved organic carbon and nitrogen in sea water. *Biol. Bull., Woods Hole*, **67**, 132.

KROGH, A., LANGE, E. and SMITH, W. (1930). On the organic matter given off by algae. *Biochem. J.* **24**, 1666.

LEVRING, T. (1945). Some culture experiments with marine plankton diatoms. *Göteborgs VetenskSamh. Handl.*, Sjätte Földjen, Ser. B, **3**, no. 12.

LEWIN, J. C. (1954). Silicon metabolism in diatoms. *J. Gen. Physiol.* **37**, 589.

LEWIN, R. A. (1954). A marine *Stichococcus* sp. which requires vitamin B$_{12}$. *J. Gen. Microbiol.* **10**, 93.

LEWIS, G. J., Jr. and THOMPSON, T. G. (1950). The effect of freezing on the sulfate/chlorinity ratio of sea water. *J. Mar. Res.* **9**, 211.

LLOYD, B. (1930). Bacteria of the Clyde sea area. *J. Mar. Biol. Ass. U.K.* **16**, 879.

LLOYD, B. (1937). Bacteria in stored sea water. *J. R. Tech. Coll. Glasg.* **4**, 173.

LOCHHEAD, A. G., BURTON, M. O. and THEXTON, R. H. (1952). A bacterial growth-factor synthesized by a soil bacterium. *Nature, Lond.,* **170**, 282.

LYMAN, J. and FLEMING, R. H. (1940). Composition of sea water. *J. Mar. Res.* **3**, 134.

MACKERETH, F. J. (1953). Phosphorus utilization by *Asterionella formosa* Hass. *J. Exp. Bot.* **4**, 296.

McCLENDON, J. F. (1917). The standardization of a new colorimetric method for the determination of the hydrogen ion concentration, CO_2 tension, and CO_2 and O_2 content of sea water. *J. Biol. Chem.* **30**, 265.

MARE, M. F. (1940). Plankton production off Plymouth. *J. Mar. Biol. Ass. U.K.* **24**, 461.

MARSHALL, S. M. and ORR, A. P. (1952). The biology of *Calanus finmarchicus*. VII. Factors affecting egg production. *J. Mar. Biol. Ass. U.K.* **30**, 527.

MATSUDAIRA, C. (1950–2). The catalytic activity of sea-water. *Tohoku J. Agric. Res.* **1**, 177; **2**, 105; **3**, 135.

MATSUE, Y. (1949). Phosphate storing by a marine plankton diatom. *J. Fish. Inst.* **2**, 29. In Japanese.

MATTHEWS, D. J. (1923). Oceanography. In *Dictionary of Applied Physics,* **3**. London.

MATTHEWS, D. J. and ELLIS, B. A. (1928). The ratio of magnesium to chlorine in the waters of the Gulf of Aden. *J. Cons. int. Explor. Mer,* **3**, 191.

MATUDAIRA, T. (1939). The physiological property of sea water considered from the effect upon the growth of diatom, with special reference to its vertical and seasonal change. *Bull. Jap. Soc. Sci. Fish.* **8**, 187.

MATUDAIRA, T. (1942). On inorganic sulphides as a growth-promoting ingredient for diatoms. *Proc. Imp. Acad. Japan,* **18**, p. 106.

MIYAKE, Y. (1939). Chemical studies of the Western Pacific Ocean *Bull. Chem. Soc. Japan,* **14**, 29, 55.

MIYAKE, Y. (1952). A table of the saturated vapour pressure of sea water. *Oceanogr. Mag.* **4**, 95.

MIYAKE, Y. and KOIZUMI, M. (1948). Measurement of the viscosity coefficient of sea water. *J. Mar. Res.* **7**, 63.

MIYAKE, Y. and SAKURAI, C. (1952). Boron in sea water as an indicator for water mass analysis. *Umi. t Sora,* **30**, no. 1. Publ. by Marine Observatory, Kobe.

MOBERG, E. G., GREENBERG, D. M., REVELLE, R. and ALLEN, E. C. (1934). The buffer mechanism of sea water. *Bull. Scripps Instn Oceanogr.* **3**, 231.

MOORE, H. B. (1931). Muds of the Clyde Sea area. III. *J. Mar. Biol. Ass. U.K.* **17**, 325.

MULLIN, J. B. and RILEY, J. P. (1954). Cadmium in sea water. *Nature, Lond.*, **174**, 42.

MUNK, W. H. and RILEY, G. A. (1952). Absorption of nutrients by aquatic plants. *J. Mar. Res.* **11**, 215.

NEWCOMBE, C. L. and BRUST, H. F. (1940). Variations in the phosphorus content of waters of Chesapeake Bay. *J. Mar. Res.* **3**, 76.

NIELSEN, E. STEEMANN (1935). The production of phytoplankton at the Faroe Isles, Iceland, East Greenland and in the waters around. *Medd. Komm. Havundersøg., Kbh.*, Ser. Plankton, **3**, no. 1.

NIELSEN, E. STEEMANN (1952). The use of radio-active carbon for measuring organic production in the sea. *J. Cons. int. Explor. Mer*, **18**, 117.

NODDACK, I. and NODDACK, W. (1940). Die Haufigkeiten der Schwermetalle in Meerestieren. *Ark. Zool.* **32** A, no. 4.

NOLL, W. (1931). Ueber die geochemische Rolle der Sorption. *Chem. d. Erde*, **6**, 552.

ORTON, J. H. (1923). Mortality among oysters. *J. Mar. Biol. Ass. U.K.* **13**, 1; Min. Agric. Fish: Fish Invest. Ser. 2, **6**, no. 3. See also Stubbs and Brady ibid no. 4.

PETERSON, W. H., FRED, E. B. and DOMOGALLA, B. P. (1925). The occurrence of amino acids and other organic nitrogen compounds in lake water. *J. Biol. Chem.* **63**, 287.

PETTERSSON, H. (1943). Manganese nodules and the chronology of the ocean floor. *Göteborgs VetenskSamh. Handl.*, Sjätte Följden, Ser. B, **2**, no. 8.

PETTERSSON, H. (1945). Iron and manganese on the ocean floor. *Göteborgs VetenskSamh. Handl.*, Sjätte Följden, Ser. B, **3**, no. 8.

POOLE, H. H. and ATKINS, W. R. G. (1937). The penetration into the sea of light of various wave-lengths as measured by emission or rectifier photo-electric cells. *Proc. Roy. Soc.* B, **123**, 151.

PROVASOLI, L. and PINTNER, I. (1953). Ecological implications of *in vitro* nutritional requirements of algal flagellates. *Ann. N.Y. Acad. Sci.* **56**, 839.

RAKESTRAW, N. W. (1936). The occurrence and significance of nitrite in the sea. *Biol. Bull., Woods Hole*, **71**, 133.

RAKESTRAW, N. W. (1947). Oxygen consumption in sea water over long periods. *J. Mar. Res.* **6**, 259.

RAKESTRAW, N. W. and EMMEL, V. M. (1937). The determination of dissolved nitrogen in water. *Industr. Engng Chem.* (Anal. ed.), **9**, 344.

RAKESTRAW, N. W. and EMMEL, V. M. (1938a). The relation of dissolved oxygen to nitrogen in some Atlantic waters. *J. Mar. Res.* **1**, 207.

RAKESTRAW, N. W. and EMMEL, V. M. (1938b). The solubility of nitrogen and argon in sea water. *J. Phys. Chem.* **42**, 1211.

RAKESTRAW, N., HERRICK, C. and URRY, W. (1939). The helium-neon content of sea water and its relation to the organic content. *J. Amer. Chem. Soc.* **61**, 2806.

RAKESTRAW, N. W. and LUTZ, F. B. (1933). Arsenic in sea water. *Biol. Bull., Woods Hole*, **65**, 397.

REDFIELD, A. C. (1934). On the proportions of organic derivatives in sea water and their relation to the composition of plankton. *James Johnstone Memorial Volume*, p. 176. Liverpool.

REDFIELD, A. C. (1942). The processes determining the concentration of oxygen, phosphate and other organic derivatives within the depth of the Atlantic Ocean. *Pap. Phys. Oceanogr.* **9**, no. 2.

REDFIELD, A. C. (1948). The exchange of oxygen across the sea surface. *J. Mar. Res.* **7**, 347.

REDFIELD, A. C. and KEYS, A. B. (1938). The distribution of ammonia in the waters of the Gulf of Maine. *Biol. Bull., Woods Hole*, **74**, 83.

REDFIELD, A. C., SMITH, H. P. and KETCHUM, B. (1937). The cycle of organic phosphorus in the Gulf of Maine. *Biol. Bull., Woods Hole*, **73**, 421.

REITH, J. F. (1930). Der Jodgehalt von Meerwasser. *Rec. Trav. chim. Pays-Bas*, **49**, 142.

RENN, C. E. (1937). Bacteria and the phosphorus cycle in the sea. *Biol. Bull., Woods Hole*, **72**, 190.

RICE, T. R. (1954). Biotic influences affecting population growth of planktonic algae. *Fish. Bull. U.S.* **87**.

RILEY, G. A. (1937). The significance of the Mississippi River drainage for biological conditions in the northern Gulf of Mexico. *J. Mar. Res.* **1**, 60.

RILEY, G. A. (1938). Plankton studies. I. *J. Mar. Res.* **1**, 335.

RILEY, G. A. (1939a). Limnological studies in Connecticut. *Ecol. Monogr.* **9**, 53.

RILEY, G. A. (1939b). Plankton studies. II. *J. Mar. Res.* **2**, 145.

RILEY, G. A. (1941a). Plankton studies. III. *Bull. Bingham Oceanogr. Coll.* **7**, art. 3.

RILEY, G. A. (1941b). Plankton studies. IV. *Bull. Bingham Oceanogr. Coll.* **7**, art 4.

RILEY, G. A. (1942). The relationship of vertical turbulence and spring diatom flowerings. *J. Mar. Res.* **5**, 67.

RILEY, G. A. (1943). Physiological aspects of spring diatom flowerings. *Bull. Bingham Oceanogr. Coll.* **8**, art. 4.

RILEY, G. A. (1946). Factors controlling phytoplankton populations on Georges Bank. *J. Mar. Res.* **6**, 54.

RILEY, G. A. (1947). A theoretical analysis of the zooplankton population of Georges Bank. *J. Mar. Res.* **6**, 104.

RILEY, G. A. (1951). Oxygen, phosphate, and nitrate in the Atlantic Ocean. *Bull. Bingham Oceanogr. Coll.* **13**, art. 1.

RILEY, G. A. and BUMPUS, D. F. (1946). Phytoplankton-zooplankton relationships on Georges Bank. *J. Mar. Res.* **6**, 33.

RILEY, G. A. and GORGY, S. (1948). Quantitative studies of summer plankton populations of the Western North Atlantic. *J. Mar. Res.* **7**, 100.

RILEY, G. A., STOMMEL, H. and BUMPUS, D. F. (1949). Quantitative ecology of the plankton of the Western North Atlantic. *Bull. Bingham Oceanogr. Coll.* **12**, art. 3.

RILEY, J. P. (1953). The spectrophotometric determination of ammonia in natural waters with particular reference to sea water. *Analyt. chim. acta*, **9**, 573.

ROBBINS, W., HERVEY, A. and STEBBIUS, E. (1950). Studies on Euglena and vitamin B_{12}. *Bull. Torrey Bot. Cl.* **77**, 423.

ROBINSON, R. A. (1954). The vapour pressure and osmotic equivalence of sea water. *J. Mar. Biol. Ass. U.K.* **33**, 449.

ROBINSON, R. J. and KNAPMAN, F. W. (1941). The sodium-chlorinity ratio of ocean waters. *J. Mar. Res.* **4**, 142.

ROBINSON, R. J. and WIRTH, H. E. (1934 *a*). Free ammonia, albuminoid nitrogen and organic nitrogen in the waters of the Puget Sound area. *J. Cons. int. Explor. Mer*, **9**, 15.

ROBINSON, R. J. and WIRTH, H. E. (1934 *b*). Free ammonia, albuminoid nitrogen and organic nitrogen in waters of the Pacific. *J. Cons. int. Explor. Mer*, **9**, 187.

ROCHE, J. and LAFON, M. (1949). On the presence of diiodotyrosine in *Laminaria*. *C.R. Acad. Sci., Paris*, **229**, 491.

RONA, E. and URRY, W. D. (1952). Radium and uranium contents of ocean and river waters. *Amer. J. Sci.* **250**, 241.

RUSSELL, F. S. (1935). The seasonal abundance of young fish in the offshore waters of the Plymouth area. *J. Mar. Biol. Ass. U.K.* **20**, 147.

RYTHER, J. H. and VACCARO, R. F. (1954). A comparison of the oxygen and ^{14}C methods of measuring marine photosynthesis. *J. Cons. int. Explor. Mer*, **20**, 25.

SARGENT, M. C. and HINDMAN, J. C. (1943). The ratio of carbon dioxide consumption to oxygen evolution in sea water in the light. *J. Mar. Res.* **5**, 131.

SCHMIDT, J. (1925). On the contents of oxygen in the ocean on both sides of Panama. *Science*, **61**, 592.

SCHREIBER, E. (1927). Die Reinkultur von marinen Phytoplankton. *Wiss. Meeresuntersuch.*, Abt. Helgoland, N.F., **16**, no. 10, pp. 1–34.

SCHULTZ, B. (1930). Die Beziehung zwischen Jod- und Salzgehalt des Meerwassers. *Ann. Hydrogr., Berl.*, **58**, 187.

SCOTT, G. T. (1945 *a*). The influence of H-ion concentration on the mineral composition of *Chlorella pyrenoidosa*. *J. Cell. Comp. Physiol.* **25**, 37.

SCOTT, G. T. (1945 *b*). The mineral composition of phosphate deficient cells of *Chlorella pyrenoidosa* during the restoration of phosphate. *J. Cell. Comp. Physiol.* **26**, 35.

SEIWELL, G. E. (1935). Note on iron analyses of Atlantic coastal waters. *Ecology*, **16**, 663.

SEIWELL, H. R. (1934). The distribution of oxygen in the western basin of the North Atlantic. *Pap. Phys. Oceanogr.* **3**, no. 1.

SEIWELL, H. R. (1935). The cycle of phosphorus in the western basin of the North Atlantic. *Pap. Phys. Oceanogr.* **3**, no 4.

SEIWELL, H. R. (1937a). The minimum oxygen concentration in the western basin of the North Atlantic. *Pap. Phys. Oceanogr.* 5, no. 3.

SEIWELL, H. R. (1937b). Consumption of oxygen in sea water under controlled laboratory conditions. *Nature, Lond.*, 140, 506.

SEIWELL, H. R. and SEIWELL, G. (1938). The sinking of decomposing plankton in sea water. *Proc. Amer. Phil. Soc.* 78, 465.

SEWELL, R. B. SEYMOUR and FAGE, L. (1948). Minimum oxygen layer in the ocean. *Nature, Lond.*, 162, 949.

SIMONS, L. H., MONAGHAN, P. H. and TAGGART, M. S. (1953). Aluminium and Iron in Atlantic and Gulf of Maine waters. *Analyt. Chem.* 25, 989.

SMALES, A. (1951). Determination of strontium in sea water by a combination of flame photometry and radio-chemistry. *Analyst*, 76, 348.

SMALES, A. A. and PATE, B. D. (1952). Determination of arsenic in sea water. *Analyst*, 77, 188.

SMITH, C. L. (1940). The Great Bahama Bank. II. *J. Mar. Res.* 3, 171.

SÖDERSTRÖM, A. (1924). *Das Problem der Polygordius-Endolarve.* 168 pp. Uppsala and Stockholm.

SPENCER, C. P. (1954). Studies on the culture of a marine diatom. *J. Mar. Biol. Ass. U.K.* 33, 265.

SPOONER, G. M. (1949). Absorption of strontium and yttrium by marine algae. *J. Mar. Biol. Ass. U.K.* 28, 587.

STANBURY, F. A. (1931). The effect of light of different intensities, reduced selectively and non-selectively, upon the rate of growth of *Nitzschia closterium*. *J. Mar. Biol. Ass. U.K.* 17, 633.

STARK, W. H., SADLER, J. and McCOY, E. (1938). Some factors affecting the bacterial population of fresh-water lakes. *J. Bact.* 36, 653.

STEPHENSON, W. (1949). Certain effects of agitation upon the release of phosphate from mud. *J. Mar. Biol. Ass. U.K.* 28, 371.

STRICKLAND, J. D. H. (1952). The preparation and properties of silico-molybdic acid. I and II. *J. Amer. Chem. Soc.* 74, 862.

SVERDRUP, H. U. (1953). On conditions for the vernal blooming of phytoplankton. *J. Cons. int. Explor. Mer*, 18, 287.

SVERDRUP, H. U., JOHNSON, M. W. and FLEMING, R. H. (1942). *The Oceans, their Physics, Chemistry and Biology.* 1087 pp. New York.

SVERDRUP, H. U. and SOULE, F. M. (1933). Scientific results of the *Nautilus* Expedition, 1931. *Pap. Phys. Oceanogr.* 2, no. 1.

TANADA, T. (1951). The photosynthetic efficiency of carotenoid pigments in *Navicula minima*. *Amer. J. Bot.* 38, 276.

THOMAS, B. D. and THOMPSON, T. G. (1933). Lithium in sea water. *Science*, 77, 547.

THOMAS, B. D., THOMPSON, T. G. and UTTERBACK, C. L. (1934). The electrical conductivity of sea water. *J. Cons. int. Explor. Mer*, 9, 28.

THOMPSON, E. F. and GILSON, H. C. (1937). Chemical and physical investigations. *John Murray Exped. Rep.* 2, no. 2.

THOMPSON, T. G. and ANDERSON, D. H. (1940). The determination of the alkalinity of sea water. *J. Mar. Res.* 3, 224.

THOMPSON, T. G. and BREMNER, R. W. (1935). The occurrence of iron in the water of the N.E. Pacific Ocean. *J. Cons. int. Explor. Mer*, 10, 39.

THOMPSON, T. G., JOHNSTON, W. R. and WIRTH, H. E. (1931). The sulphate-chlorinity ratio in ocean waters. *J. Cons. int. Explor. Mer*, 6, 246.

THOMPSON, T. G. and KORPI, E. (1942). The bromine-chlorinity ratio of sea water. *J. Mar. Res.* 5, 28.

THOMPSON, T. G., LANG, J. W. and ANDERSON, L. (1927). The sulfate-chloride ratio of the waters of the North Pacific. *Publ. Puget Sd Mar. (Biol.) Sta.* 5, 277.

THOMPSON, T. G. and ROBINSON, R. (1932). Chemistry of the sea. *Bull. Nat. Res. Coun., Wash.*, no. 85.

THOMPSON, T. G. and TAYLOR, H. J. (1933). Determination and occurrence of fluorides in sea water. *Industr. Engng Chem.* (Anal. ed.), 5, 87.

THOMPSON, T. G. and WILSON, T. L. (1935). The occurrence and determination of manganese in sea water. *J. Amer. Chem. Soc.* 57, 233.

THOMPSON, T. G. and WRIGHT, C. C. (1930). Ionic ratios of the waters of the North Pacific Ocean. *J. Amer. Chem. Soc.* 52, 915.

THOMSEN, H. (1948). Instructions pratiques sur la détermination de la salinité de l'eau de mer. *Bull. Inst. océanogr. Monaco*, no. 930.

UTTERBACK, C. L., THOMPSON, T. G. and THOMAS, B. D. (1934). Refractivity-chlorinity-temperature relationships of ocean waters. *J. Cons. int. Explor. Mer*, 9, 35.

VACCARO, R. F. and RYTHER, J. H. (1954). The bactericidal effects of sunlight in relation to 'light' and 'dark' bottle photosynthesis experiments. *J. Cons. int. Explor. Mer*, 20, 18.

VACCARO, R. F., BRIGGS, M. P., CAREY, C. L. and KETCHUM, B. H. (1950). Viability of *Escherichia coli* in sea water. *Amer. J. Publ. Hlth*, 40, 1257.

VINOGRADOV, A. P. (1953). The elementary composition of marine organisms. *Mem. II, Sears Found. Mar. Res.* New Haven, Conn.

WAKSMAN, S. A. and CAREY, C. L. (1935). Decomposition of organic matter in sea water by bacteria. I and II. *J. Bact.* 29, 531.

WAKSMAN, S. A., CAREY, C. L. and REUSZER, H. W. (1933). Marine bacteria and their role in the cycle of life in the sea. I. Decomposition of marine plant and animal residues by bacteria. *Biol. Bull., Woods Hole*, 65, 57.

WAKSMAN, S. A. and HOTCHKISS, M. (1937). Viability of bacteria in sea water. *J. Bact.* 33, 389.

WAKSMAN, S. A., HOTCHKISS, M. and CAREY, C. L. (1933). Bacteria concerned in the cycle of nitrogen in the sea. *Biol. Bull., Woods Hole*, 65, 137.

WAKSMAN, S. A., HOTCHKISS, M., CAREY, C. L. and HARDMAN, Y. (1938). Decomposition of nitrogenous substances in sea water by bacteria. *J. Bact.* 35, 477.

WAKSMAN, S. A. and RENN, C. E. (1936). Decomposition of organic matter in sea water. III. *Biol. Bull., Woods Hole*, **70**, 472.

WAKSMAN, S. A., REUSZER, H. W., CAREY, C. L., HOTCHKISS, M. and RENN, C. E. (1933). Biology and chemistry of the Gulf of Maine. III. *Biol. Bull., Woods Hole*, **64**, 183.

WAKSMAN, S. A., STOKES, J. L. and BUTLER, M. R. (1937). Relation of bacteria to diatoms in sea water. *J. Mar. Biol. Ass. U.K.* **22**, 359.

WASSINCK, E. C. and KERSTEN, J. A. H. (1944). Observations sur la photosynthèse et la fluorescence chlorophyllienne des diatomées. *Enzymologia*, **11**, 282.

WATTENBERG, H. (1929). Die Durchluftung des Atlantischen Ozeans. *J. Cons. int. Explor. Mer*, **4**, 68.

WATTENBERG, H. (1933). Kalzium Karbonat- und Kohlensäuregehalt des Meerwassers. *Wiss. Ergebn. dtsch. atlant. Exped. 'Meteor'*, **8**.

WATTENBERG, H. (1937a). Die Bedeutung anorganischer Factoren bei Ablagerung von Kalzium Karbonat im Meere. *Geol. Meere*, **1**, 237.

WATTENBERG, H. (1937b). Methoden zur Bestimmung von Phosphat, Silikat, Nitrit, Nitrat und Ammoniak im Seewasser. *Rapp. Cons. Explor. Mer*, **103**, 1.

WATTENBERG, H. (1937c). Die chemischen Arbeiten auf der 'Meteor'-Fahrt Feb.–Mai 1937. *Ann. Hydrogr., Berl.*, usw. Sept. Beiheft, p. 17.

WATTENBERG, H. (1938). Zur Chemie des Meerwassers. *Z. anorg. Chem.* **236**, 339.

WATTENBERG, H. and TIMMERMANN, E. (1936). Ueber die Sättigung des Seewassers an CaCO₃. *Ann. Hydrogr., Berl.*, usw. p. 23.

WATTENBERG, H. and TIMMERMANN, E. (1937). Die Löslichkeit von Magnesiumkarbonat und Strontiumkarbonat in Seewasser. *Kieler Meeresforsch.* **2**, 81.

WEBB, D. A. (1937). Spectrographic analyses of marine invertebrates. *Sci. Proc. R. Dublin Soc.* **21**, 505.

WEBB, D. A. (1938). Strontium in sea water and its effect on calcium determinations. *Nature, Lond.*, **142**, 751.

WENNER, F., SMITH, E. H. and SOULE, F. M. (1930). Apparatus for the determination aboard ship of the salinity of sea water by the electrical conductivity method. *J. Res. Nat. Bur. Stand.* **5**, 711.

WHIPPLE, G. C. (1901). Changes that take place in the bacterial contents of waters during transportation. *Technol. Quart.* **14**, 21.

WIBORG, K. F. (1948). Investigations on cod larvae in the coastal waters of Northern Norway. *Fiskeridir. Skr. Havundersøk.* **9**, no. 3.

WILSON, D. P. (1932). On the mitraria larva of *Owenia fusiformis* Delle Chiaje. *Phil. Trans.* B, **221**, 231.

WILSON, D. P. (1951). A biological difference between natural sea waters. *J. Mar. Biol. Ass. U.K.* **30**, 1.

WOOD, E. J. F. (1951). Bacteria in marine environments. *Proc. Indo-Pacif. Fish. Coun.*, 2nd Meeting, 1950, Sect. II, p. 69.

WOOD, E. J. F. (1953). Heterotrophic bacteria in marine environment of Eastern Australia. *Aust. J. Mar. Freshw. Res.* **4**, 160.

ZOBELL, C. E. (1935a). The assimilation of ammonium nitrogen by marine phytoplankton. *Proc. Nat. Acad. Sci., Wash.*, **21**, 517.

ZOBELL, C. E. (1935b). Oxidation-reduction potentials and the activity of marine nitrifiers. *J. Bact.* **29**, 78.

ZOBELL, C. E. (1936). Bactericidal action of sea-water. *Proc. Soc. Exp. Biol., N.Y.*, **34**, 113.

ZOBELL, C. E. (1940a). Some factors which influence oxygen consumption by bacteria in lake water. *Biol. Bull., Woods Hole*, **78**, 388.

ZOBELL, C. E. (1940b). The effect of oxygen tension on the rate of oxidation of organic matter in sea water by bacteria. *J. Mar. Res.* **3**, 211.

ZOBELL, C. E. (1941). Studies on marine bacteria. I. *J. Mar. Res.* **4**, 42.

ZOBELL, C. E. (1946). *Marine Microbiology*. Waltham, Mass.

ZOBELL, C. E. and ANDERSON, D. Q. (1936a). Observations on the multiplication of bacteria in different volumes of stored sea water. *Biol. Bull., Woods Hole*, **71**, 324.

ZOBELL, C. E. and ANDERSON, D. Q. (1936b). Vertical distribution of bacteria in marine sediments. *Bull. Amer. Ass. Petrol. Geol.* **20**, 260.

ZOBELL, C. E. and FELTHAM, C. B. (1935). The occurrence and activity of urea-splitting bacteria in the sea. *Science*, **81**, 234.

ZOBELL, C. E. and UPHAM, H. C. (1944). A list of marine bacteria, etc. *Bull. Scripps Inst. Oceanogr.* **5**, 239.

ZWICKER, B. M. G. and ROBINSON, R. J. (1944). The photometric determination of nitrate in sea water with a strychnidine reagent. *J. Mar. Res.* **5**, 214.

ADDENDA

THE following notes and references to recent publications
expand the information relating to the environment of marine
plants and animals which are included in the text.

APRIL 1959

PAGE 26

Krey has reviewed the various methods which have been used
to estimate the standing crop of phytoplankton from the chlorophyll present in the residue filtered from samples of water
(*Rapp. Cons. Explor. Mer*, **144**, 20, 1958).

A method of estimating each of the three chlorophylls, *a*, *b*
and *c*, when together in an acetone extract of samples of
plankton, by Richards and Thompson (*J. Mar. Res.* **11**, 156,
1952) has been used extensively.

PAGE 28

The mineral nutrition of phytoplankton has been reviewed by
Ketchum in *Annu. Rev. Pl. Physiol.* 1954.

PAGE 35

A marked inverse relation has been found between the organic
content of marine sediments and the oxygen content of the overlying water in the Gulf of Mexico (Richards and Redfield. *Deep
Sea Res.* **1**, 279, 1954).

This and later investigations are included in a review entitled
'Oxygen in the Oceans' by Richards (*Mem. Geol. Soc. Amer.*
no. 67, vol. 1, pp. 185–238, 1957).

PAGE 40

A study of remineralization of organic phosphorus in plankton
has been made by Hoffmann (*Kieler Meeresforsch.* **12**, 25, 1956).

PAGE 41

In the inner part of the Gulf of Venezuela, the deep water,
cut off from the ocean by a sill, was found to be notably rich in
organic phosphorus. This is attributed to organisms and particulate matter sinking into the basin and carrying phosphate down
from the surface layers (Redfield. *Deep Sea Res.* **3**, part 2,
p. 115, 1955).

PAGE 42

Buljan (*Bull. volcan.* Serie 2, vol. 17, p. 41, 1955) attributes the high phosphate content of the deep water in the Tyrrhenian Basin, encompassed by land or shallow submarine sills, to enrichment by volcanic action. He further discusses the possibility that the greater concentrations of phosphate in the Caribbean and in the Pacific than in the Atlantic may be due to the same cause.

PAGE 43

Observations concerning the sorption of phosphate in suspended solids are discussed by Carritt and Goodgal (*Deep Sea Res.* **1**, 224, 1954).

A study of the regeneration of phosphate, ammonia and silicate in three deep basins off the California coast has been made by Rittenberg, Emery and Orr (*Deep Sea Res.* **3**, 23, 1954).

PAGE 46

In the equatorial Atlantic, numerous surface samples showed significant concentrations of organic phosphorus, whereas below 1000 m. none was detected (Ketchum, Corwin and Keen. *Deep Sea Res.* **2**, 172, 1955).

PAGE 54

From cores collected in the equatorial Atlantic, Wiseman estimates that the rate of accumulation of organic nitrogen on the sea floor averages some 3 mg. N per m.^{2} annually (*Rep. Brit. Ass.* 1956, p. 579).

PAGE 61

The bacteria present in an offshore area have been found by Huddleston (*J. Appl. Bact.* **18**, 22, 1955) to consist mainly of obligate halophytes, 96 % being small gram-negative rods. The bacterial content of the water was greatest in the summer and showed little significant variation with depth.

PAGE 77

Experiment has shown that bacterial oxidation of added ammonia does not occur in samples of offshore sea waters, unless ferric hydroxide is added, which may be in the form of a ferric salt or as sterile diatoms on which ferric hydroxide has accumulated (Spencer. *J. Mar. Biol. Ass. U.K.* **35**, 621, 1956).

PAGE 85

Measurements of illumination, expressed in terms of light energy, down to great depths in the ocean have been made by Clarke and Wertheim (*Deep Sea Res.* **3**, 189, 1956) and include measurements at night of light due to luminescence by organisms.

PAGE 86

There is growing evidence that the 'physiological state' of phytoplankton, that is the rate at which the cells can photosynthesize when illuminated for a few hours, changes considerably throughout the day and night. This change depends upon the light intensity to which the cells have been exposed previously, the duration of exposure and the temperature. It appears to be linked with the cells' changing chlorophyll content, which can increase during darkness or dim light, although probably not wholly dependent upon their chlorophyll content.

An extensive investigation has been made of the exponential growth rate of the fresh-water alga *Chlorella ellipsoidea* at 10, 2 and 0·4 kilolux, 25°, 15° and 7° C., in continuous light and with periods of 6, 12 and 18 hours' darkness daily (Tamiya *et al. Biochim. biophys. Acta*, **12**, 23, 1953; *J. Gen. Appl. Microbiol.* **1**, 298, 1955).

When grown at 25° C. in continuous bright light and transferred to darkness, the chlorophyll content of the cells per unit dry weight increased during the period of darkness. In this they resembled the brown alga Phaeodactylum when nitrogen deficient (Harvey, 1953b) and also, as found in recent experiments, when grown with ample nitrate in enriched sea water.

When grown in 0·4 kilolux a period of darkness had no effect on the subsequent growth rate of the *Chlorella* but when grown in 10 or 2 kilolux at 15° or 7°, even a 6-hour period of darkness caused a considerable increase in their exponential growth rate when again illuminated. This increase was greater at 7° than at 15° C. Indeed, when illuminated at 7° C. in light of 10 kilolux for 12 hours with intermittent periods of 12 hours' darkness, the crop was greater than when grown in continuous light of the same intensity. The changes taking place during darkness had a remarkably large temperature coefficient compared with that of the changes taking place while illuminated. Growth rate was measured by change in packed cell volume, which had been found to be proportional to dry weight.

Experiments with Phaeodactylum (unpublished) growing exponentially with ample nutrients in the sea water at 16° and 8 kilolux, in which CO_2 assimilation was measured by change in pH of the sea water, have shown marked increases in exponential growth rate after a period in the dark. In similar experiments with Phaeodactylum growing exponentially at 16° C. and at 5° C., the chlorophyll content of the cells increased by around 25 % during periods of 18 hours' darkness.

These indications of analogous response to periods of darkness between the fresh-water and marine alga, suggest a metabolic mechanism which materially helps the growth of phytoplankton in nature.

Doty and Oguri (*Limnol. & Oceanogr.* **2**, 37, 1957) collected many surface samples in the tropical Pacific at 8 a.m. and 7 p.m., enriched them with $NaH^{14}CO_3$ and exposed them to constant light at sea temperature, for 4 or 6 hours. The uptake of ^{14}C was severalfold greater in those collected at 8 a.m. than in those collected at 7 p.m. Similar diurnal changes in the uptake of ^{14}C by the phytoplankton took place in samples taken at intervals during a 24-hour period from a carboy of surface water, lashed on deck.

PAGE 87

This once accepted conclusion that 22·4 cm.³ difference in oxygen between the light and dark bottles approximates to the photosynthesis of 12 mg. C is not strictly valid. In the first place the CO_2 is converted into a mixture of protein, lipid and carbohydrate which indicates a photosynthesis of about $0·84 \times 12$ mg. C under normal circumstances (p. 30). If lipid only were being laid down, which might almost happen for short periods in nitrogen-deficient cells, as little as 0·7 mg. C might be photosynthesized.

In the second place it is not certain that, under all circumstances, the rate of respiration in the dark is the same as in the light—the building up of end products in the cells does not cease suddenly when they are put into the dark; if nitrogen deficient they even continue to utilize nitrate or ammonium to build up organic nitrogen compounds. In consequence a gradual diminution of respiratory loss will take place when cultures are transferred from the light to the dark as they adapt their basic metabolism to the changed conditions.

Indeed there seem to be *prima facie* reasons to doubt whether either the photosynthetic quotient or rate of loss by respiration are not affected by conditions of micronutrient supply, particularly nitrogen supply.

PAGE 88

Experiments with cultures growing in different light intensities by Ryther have shown optimum growth (measured by ^{14}C uptake) for diatoms in the range of 10–25 kilolux, for dinoflagellates in 25–35 kilolux and for Chlorophyta within the range of 5–10 kilolux (*Limnol. & Oceanogr.* **1**, 61, 1956).

PAGE 89

Observations linking increase in cellular carbon deduced from ^{14}C experiments with oxygen production have been made by Ryther (*Deep Sea Res.* **2**, 134, 1955) and by Steeman Nielsen (*Physiol. Plant.* **9**, 144, 1956).

During short periods of illumination, or at near light saturation and when the algae are not deficient in N, P or other necessary micronutrients, it appears that the newly formed labelled tissue is not broken down and respired. It is not until later that this new tissue takes part in respiratory breakdown.

In cells which are deficient, notably in nitrogen, the rate of loss by respiration is considerable in comparison with that of the slowed rate of photosynthesis. With such deficient cells it appears that during short periods of illumination the increase in cellular carbon deduced from ^{14}C uptake is much less than the gross production by photosynthesis, unless a very low photosynthetic quotient is used to calculate photosynthesis from the O_2 production (*vide supra*).

The *exact* interpretation of values found for increase in cellular carbon from ^{14}C experiments in terms of absolute quantities of carbon which has been photosynthesized appears to be unsolved as yet. For a review of the present situation see Ketchum *et al.* (*Rapp. Cons. Explor. Mer*, **144**, 132, 1958) and Steeman Nielsen (*ibid.* **144**, 385, 1958).

In experiments quoted by Ketchum *et al.* with nutrient-poor

Surface water exposed for 3 days to natural light

Gross production (from O_2 production using quotient $\dfrac{CO_2}{O_2} = 0.75$)	Net production (from ^{14}C uptake)
No enrichment 0·0125 Enriched N and P 0·135	0·0018 (g. $C/m.^2/day$) 0·11 (g. $C/m.^2/day$)

raw surface waters from the Caribbean, the effect of adding nitrate and phosphate was usually very marked, but with some waters this had little or no effect. This suggests a lack of micronutrients other than N and P occurring at times in nature.

The addition of nutrients reduced the loss by respiration from 80 to 18% of the gross production in this experiment.

PAGE 94

A series of investigations, has recently been made concerning the relation between pigment content of the phytoplankton and organic production in culture and also in the sea.

In low intensities the light absorbed and supplying energy for photosynthesis by chlorophyll a is reinforced by light, energy absorbed by chlorophylls b, c and fucoxanthin. With increasing illumination an intensity is reached above which this extra supply of light energy from the accessory pigments can no longer be utilized by the chlorophyll a, which itself is absorbing all the light energy which it can use for photosynthesis. Hence a rather close relation would be expected between the chlorophyll a present in phytoplankton and its rate of photosynthesis when light saturated, provided the plants are in a 'physiological state' permitting maximum photosynthesis to take place when they are light saturated. This expectation may be expanded—the 'physiological state' permitting maximum assimilation is likely to depend upon a sufficient initial content and a sufficient rate of supply to the cells during growth of N, P, Fe, Mn (and of SiO_2 and B_{12} to diatoms), also upon there not being partial exhaustion of necessary intermediates, such as enzymes, which may have taken place during previous growth in moderate or strong light and not have had time to build up again during darkness or dim light.

This expectation was borne out in a recent experiment in which eighteen species of phytoplankton algae, belonging to a range of taxonomic groups with differing proportions of accessory pigments, in unialgal culture, were exposed to artificial light for six hours and their oxygen production determined. The ratio of oxygen produced to mg. of chlorophyll a per litre of culture was nearly the same except for three species of flagellates (Ryther, *Limnol. & Oceanogr.* **1**, 72, 1956).

In the sea most of the plant production takes place at depths with intensities less than light saturation (fig. 41, p. 88). Moreover both species, and their relative proportions of accessory pigments, may vary widely from area to area, as on passing from a diatom dominant population to a mainly coccolithophore population. In the upper layers during high summer and in the tropics the natural phytoplankton is likely to be nitrogen deficient, and nitrogen-deficient cells have an impaired growth rate and have been found to contain a lesser ratio of chlorophyll a to accessory pigments (Yentsch and Vaccarro, *Limnol. & Oceanogr.* **3**, 443, 1958).

Raw waters containing a natural population from a number of positions have been illuminated at light saturation and the mean value of the ratio mg. C synthesized hourly per mg. chlorophyll a has been arrived at. This mean value in conjunction with the mean daily solar radiation, the extinction coefficient of light absorption in the water and the chlorophyll a content of the suspended plants (and detritus) was used by Ryther and Yentsch (*Limnol. & Oceanogr.* **2**, 281, 1957; **3**, 327, 1958) to calculate the average daily production of organic carbon below a square metre of sea surface. Production values calculated in this manner compared reasonably with values obtained by *in situ* productivity methods. These and other observations bearing upon the ratio of assimilation to chlorophyll a content and upon the interpretation of values for ^{14}C uptake are reviewed by Ketchum, Ryther, Yentsch and Corwin (*Rapp. Cons. Explor. Mer*, **144**, 134, 1958). Currie found a closer relation between *in situ* measured production and total pigments than between production and chlorophyll a, when com-

224 ADDENDA

paring diatom rich waters off the Portuguese coast with waters 200 miles to seaward containing less phytoplankton, mainly coccolithophores relatively rich in chlorophyll c (*Rapp. Cons. Explor. Mer*, **144**, 96, 1958).

PAGE 97

Ohle presents the results of experiments in lake waters of low nutrient content which indicate that the 'washing effect' of turbulent motion has a very marked influence on growth rate, due to the renewal of the contact layer at the surface of the plant cells.

PAGE 99

It should now be possible, by measuring production of organic carbon by phytoplankton using the ^{14}C technique, to determine whether the addition of iron to ocean waters far from land increases the growth rate of its natural population.

PAGE 101

The depth of the wind-produced homogeneous layer in the oceans throughout the year, that is the layer of greatest vertical eddy motion affecting the plants, has been studied by Lumby (*Fish. Invest. Lond.* Series 2, vol. 20, no. 2, 1955).

PAGE 103

Observations by Lewin and by Lewin and Philpott (*J. Gen. Microbiol.* **18**, 418, 1958; **18**, 427, 1958) have shown that cells which have changed from spindle or triradiate form into oval form then possess one single silicate valve resembling the valves of the diatom *Cymbella*.

PAGE 107

Evidence is accumulating that dissolved vitamin B_{12} or its analogues undergo a seasonal variation in the sea, that it is collected by both animals and plants, and is more concentrated in inshore than in oceanic surface water, that much occurs in marine mud, and that many species of marine bacteria produce it.

Cowey (*J. Mar. Biol. Ass. U.K.* **35**, 609, 1956) found by bioassay some 2μg. per m.^{3} in sterilized North Sea surface waters in the spring and $0\cdot2\mu$g. per m.^{3} during summer, while surface waters from far out in the Atlantic contained some $0\cdot1\mu$g. per m.^{3}. It is noteworthy that these waters contained more B_{12}

after storage without sterilizing preservative, presumably due to its formation by proliferating bacteria.

These concentrations compare with some 10μg. per m.[3] found by Lewin off Nova Scotia during winter and 5–10μg. per m.[3] found in inshore Scottish water during winter by Droop ($J. Mar. Biol. Ass. U.K.$ **34**, 435, 1955).

Droop has also found that the inshore diatom, $Skeletonema$ $costatum$, in bacteria-free culture required B_{12} for continued growth, and that the quantity required could be spared by pseudo-vitamin B_{12}, factor A or factor B. The flagellate $Monochrysis$ $lutheri$ required B_{12}, pseudovitamin B_{12} or factor A, but could not use factor B ($J. Mar. Biol. Ass. U.K.$ **34**, 224 and 435, 1955).

Daisley ($J. Mar. Biol. Ass. U.K.$ **37**, 683, 1958) has made observations at positions in the Bay of Biscay showing an average of 0.57μg./m.[3] B_{12} in the waters above 190 m. and below 2110 m. depth, with an average of 2.26μg./m.[3] at intermediate depths.

Thus present knowledge indicates that nearly all species of planktonic unicellular algae have an absolute requirement for B_{12}, and that sea water often contains insufficient for more than a quite limited growth. But there is no evidence as yet that growth *rate* of natural populations is impaired by the low concentrations existing in the upper layers of oceanic water. Natural population in this concept must be held to include bacteria always associated with phytoplankton, and perhaps the zooplankton grazing upon the plants. As with the possible lack of available iron, it should now be possible to deduce whether primary production of organic matter by the natural populations is impaired by low B_{12} concentrations in the sea, by means of the ^{14}C technique.

Lundin and Ericson (*2nd Internat. Seaweed Symp.* 1956, Pergamon Press) found that twenty-four out of thirty-four species of marine bacteria examined produced B_{12}; that seaweeds contained material quantities and deduced that this is absorbed by the weeds from B_{12} produced by attached bacteria.

Burkholder and Burkholder (*Science*, **123**, 1071, 1956) extracted 2.5μg. B_{12} from 1 gm. of estuarine mud, while Cowey (*loc. cit.*) has found a fraction of a microgram per gram wet weight in various zoo- and phytoplankton.

PAGE 109

The absorption of silicate and deposition of silica in the frustules of *Navicula pellicosa* has been studied further by Mrs Lewin. Colloidal silica, which does not react with molybdate, could not be utilized (*Plant Physiol.* **30**, 129, 1955). A number of compounds, which do not contain sulphur, stimulated the utilization of orthosilicate. All of these, including glucose, lactate, citrate and glycerol, also stimulated respiration (*J. Gen. Physiol.* **39**, 1, 1955).

These observations indicate that the deposition of silica in diatom frustules is not necessarily linked with their sulphur metabolism.

PAGE 110

Spencer (*J. Mar. Biol. Ass. U.K.* **37**, 127, 1958) has investigated the ionic concentrations of copper, manganese and zinc when added to sea water containing varying concentrations of ethylenediamine tetra-acetate.

The complex with iron, stable in the dark, is decomposed yielding ferric hydroxide in the light.

PAGE 118

On feeding female *Calanus* with several species of phytoplankton, labelled with ^{14}C or with ^{32}P, some 50–80 % of the radioactivity was retained by the *Calanus*, the major part of the algal organic matter being digested and built up into the animal, much of it into the developing eggs (Marshall and Orr. *Deep Sea Res.* **3**, part 2, p. 110, 1955 and *J. Mar. Biol. Ass. U.K.* **34**, 495, 1955).

PAGE 120

An explanation of variations in nutrient content and biological productivity in the English Channel has been sought in terms of the influence of Arctic climate on deep water movements in the north east Atlantic by Cooper (*Deep Sea Res.* **3**, part 2, p. 212 and *J. Mar. Res.* **14**, 347, 1955).

PAGE 122

Three sea areas have been recently surveyed at intervals with the aim of connecting the environmental condition with their productivity.

An extensive 'patch' of water, richer in plankton than the surrounding sea, drifting southward in the North Sea, has been followed and contoured at intervals between April and June for phytoplankton, zooplankton and pelagic fish. An echo survey allowed the density of pelagic fish—herring—to be assessed. The decrease in cell width of the dominant diatoms, and in their cell wall thickness as measured with an electron microscope, provided an estimate of their division rate; changes in phosphate content of the water were used to estimate the production of organic matter during each interval. The relations found between the diatoms, the herbivorous zooplankton and herring are presented in a monograph by Cushing (*Fish. Invest. Series* 2, Lond. vol. 18, no. 7, 1955). The following are amongst the conclusions: a production of about one-third of the standing crop of diatoms occurred daily in late April–early May; there was a big difference in division rate between different species of diatoms; a considerable natural mortality of diatoms occurred, many unbroken empty frustules being found; of the *Calanus*, which constituted much of the zooplankton population, the greatest mortality took place while in Stages IV and V; the number of adults were small until the pelagic fish disengaged from the 'patch', and then they were quadrupled; as the available plant food became reduced, recruitment to the earlier stages was reduced.

The waters of Long Island Sound have been surveyed at monthly intervals throughout two years (Riley. *Deep Sea Res.* 3, part 2, p. 224, 1955). The major limiting factors for plant production were the available nitrogen supply in spring and summer, and the effective light in autumn and winter. More available nitrogen, phytoplankton and zooplankton was found during one of the two summers. Most of the plant production is utilized in support of small animals with high respiration rates which do not provide sufficient food for large carnivora, but the area (average depth 20 m.) supports a large biomass of bottom-living animals. A comparison is made with the productivity of the English Channel.

A series of positions on the Fladen grounds, lying between Scotland and Norway and having a depth of 150 m. has been surveyed at intervals throughout several years. The area is but little subject to horizontal currents. An aim has been to connect the production of plant organic phosphorus and carbon during intervals between cruises with the factors influencing production, and, finally, to construct a 'mathematical model' connecting changes in the standing crop of phytoplankton. Two papers have been published (Steele. *J. Mar. Biol. Ass. U.K.* **35**, 1, 1956 and **36**, 233, 1957), and a review is in preparation for publication in *Biological Reviews*.

Production in terms of organic carbon in successive layers between cruises has been calculated from the phosphate taken up and the $C:P$ ratio in the phytoplankton, and also from ^{14}C measurements. The phosphate taken up by the plants—the 'biological change in phosphate'—was derived from the change in phosphate content of the water in each layer after adjusting (i) for regeneration, which was assumed equal throughout the water column to the regeneration which had taken place in the deep water, and (ii) for refreshment from below by vertical mixing. The latter was derived from the heat exchange observed in the water column. The observations showed that the zone of phosphate utilization and greatest chlorophyll content sank progressively during the summer to a depth of 30–40 m. The calculations indicated an annual production of plants containing $c.$ 80 gm. C below a square metre.

PAGES 126 and 131

Evidence has been presented by Baas Becking (*Nature, Lond.*, **182**, 645–647, 1958) which provides strong suspicion that sea waters may contain significant quantities of very stable perchlorate ions. These are not precipitated by silver nitrate. The quantities indicated are some 0·10 gm. $ClCO_4$ per litre in surface off-shore water and 0·03 g. in deeper water, while in inshore waters quantities up to 0·3 g. per litre were indicated. This throws doubt upon the constancy of the relation between chlorinity or salinity and density.

PAGE 130

Wilson and Arons (*J. Mar. Res.* **14**, 195, 1955) have made vapour pressure determinations and computed osmotic pressures between 0 and 35° C. and between 0 and 30 ‰ chlorinities.

PAGE 135

The calcium/strontium ratio and the organic matter in the calcareous portions of a large number of marine organisms has been investigated by Thompson and Chow (*Deep Sea Res.* **3**, part 2, p. 29, 1955). The ratio is much smaller in recent than in fossil corals (Bowen. *J. Mar. Biol. Ass. U.K.* **35**, 451, 1956). The uptake and exchange of strontium by twelve species of planktonic algae has been examined by Rice (*Limnol. & Oceanogr.* **1**, 123, 1956). The uptake by *Carteria* was dependent upon the physiological state of the cells and was independent of the concentration of calcium.

PAGE 140, Table 17

The following analyses of microconstituents in samples of sea water have been published recently:

	mg. per m.³	Reference
Barium	6	Bowen. *J. Mar. Biol. Ass. U.K.* **35**, 451, 1956
Caesium	0·5	Smales and Salmon. *Analyst*, **80**, 37, 1955
Chromium	0·04–0·07	Ishibashi. *Rec. oceanogr. Wks Jap.* **1**, 88, 1953
Cobalt	0·38–0·67	Ishibashi, *loc. cit.*
Gold	0·5–0·015	Hummel. *Analyst*, in press
Mercury	0·03	Stock and Cucnel. *Naturwiss.* **22**, 390, 1934
Molybdenum	9–11	Ishibashi, *loc. cit.*
Nickel	1·4–2·6	Levaster and Thompson. *J. Cons. int. Explor. Mer*, **21**, 125, 1956
Nickel	0·7–0·8	Ishibashi, *loc. cit.*
Rubidium	120	Smales and Salmon. *Analyst*, **80**, 37, 1955
Rubidium	120 ± 10	Smales and Webster. *Geochim. et Geophys. Acta*, **11**, 139, 1957
Selenium	4–6	Ishibashi, *loc. cit.*
Tungsten	0·09–0·10	Ishibashi, *loc. cit.*
Uranium	2·5	Stewart and Bently. *Science*, **120**, 50, 1954
Vanadium	2·6–4·6	Ishibashi, *loc. cit.*

Few of the analyses quoted are of deep-water samples, some are of inshore waters, and some are the results of single analyses. In addition to errors inherent in analytical techniques, adsorption of some elements on epiphytic bacteria developing on the walls of the container during storage may account for some low values.

In the upper layers of the sea there is adsorption on and absorption by plankton, and adsorption on inorganic particles in suspension, of at least some elements; sinking will result. The ultimate fate, whether re-solution or sedimentation, of micro-constituents other than ionic nitrogen compounds and phosphate which have been collected by plankton organisms has not been investigated. Material quantities of some, such as titanium, iron and manganese, are deposited on the ocean floor. Refreshment of inshore waters by rivers varies with the geological formation of the catchment areas.

Krauskopf (*Geochim. et Cosmochim. Acta*, **9**, 1–12 B, 1956) has found thirteen metal microconstituents to be undersaturated in sea water. Adsorption on particles has been investigated, absorption by organisms and localized formation of insoluble sulphide have been considered as possible causes.

Page 147 a

Mercury. When a dilute solution of mercuric chloride in raw sea water (about 1 mg. Hg per litre) is stored, there is a loss of mercury in solution due to its being taken up by bacteria, and subsequently being converted in part to a volatile compound. The loss does not take place in sterile sea water (Corner and Rigler. *J. Mar. Biol. Ass. U.K.* in the press).

Page 147 b

About half the iodine in solution in sea water is immediately precipitated by silver nitrate, the remainder being more likely hypoiodite than iodate (Cooper and Shaw. *J. Mar. Biol. Ass. U.K.* in the press).

It is of interest that exposed sea weed gives off molecular iodine; that the sea is likely to contain traces of iodine in equilibrium with other oxidized forms of iodide; and that air from the sea contains more iodine than air from inland (cf. *Sea-weeds and their uses*, Chapman, 1950; *Iodine and the incidence of goitre*, McClendon, 1939).

Page 149

A modification of Krogh's technique has been developed by Kay for serial analyses of organic carbon in sea water, and values ranging around 3 mg. C per litre found at various positions in Kiel Bay, with a maximum at 10 m. depth and least in water near the bottom (*Kieler Meeresforsch*. **10**, 26. 1954; **10**, 202, 1954).

Waters from three positions off the California coast have been analysed by Plunkett and Rakestraw (*Deep Sea Res.* **3**, part 2, p. 12, 1955). About $2\frac{1}{2}$ mg. C per litre was found in the upper layers and at 1000 m. and below, with minimum concentrations of about $1\frac{1}{2}$ mg. C per litre at depths of 500–700 m. In two stations in the west Pacific concentrations ranging around $1\frac{1}{2}$ mg. C per litre were found down to a depth of 6000 m.

These various observations (see also p. 71) suggest that bacteria are unable to cause further breakdown of organic matter when its concentration falls below about 1 mg. C per litre or alternatively that this final residue of organic matter is remarkably resistant to bacterial attack.

PAGE 151

A review of present knowledge concerning organic compounds which have been found dissolved in both sea and fresh waters has been published by Vallentyne (*J. Fish. Res. Bd Can.* **14**, 33, 1957).

PAGE 160

A method of estimating 'Titration Alkalinity', by adding acid to a sample of water and determining the change in pH, evolved by Anderson and Robinson (*Industr. Engng Chem. (Anal.)*, **18**, 767, 1946), has been used extensively by Koczy (*Deep Sea Res.* **3**, 279, 1956) to determine the 'Specific Alkalinity' in equatorial waters of the Atlantic, Pacific and Indian Oceans, and its variation with depth. A notable increase was found in waters with temperatures below about 9° C.

PAGE 183

Richards and Corwin (*Limnol. & Oceanogr.* **1**, 263, 1956) have published a useful monograph giving the relation between concentrations of O_2 in equilibrium with the atmosphere and salinity and temperature, based on Truesdale and Downing's observations.

PAGE 195

A method is described of storing water samples for subsequent estimation of total phosphorus, in which bacteria, which have grown attached to the walls of 'Duraglass' storage bottles, are rubbed off with a policeman (Ketchum, Corwin and Keen. *Deep Sea Res.* **2**, 172, 1955).

PAGES 196–8

A method has been developed by Riley and Sinhaseni (*J. Mar. Biol. Ass. U.K.* **36**, 161, 1957) for the determination of total NH_4-N, NO_2-N and NO_3-N in sea waters. The nitrate and nitrite in a 50 ml. sample are wholly reduced to ammonia with 'Raney nickel' in the presence of disodium ethylenediamine tetra-acetate. The sample is made alkaline and the evolved ammonia is collected in a small volume of dilute acid by a convenient diffusion method which gives a 73% recovery. Alternatively it may be distilled under reduced pressure which gives a 100% recovery (see *Analyt. chim. acta*, **9**, 573, 1953). The ammonia in the distillate is reacted with an alkaline phenate solution containing methyl alcohol and acetone followed by a solution of hypochlorite. The blue indophenol develops in the cold and is determined photometrically. The method is very sensitive, and no interference was found from a variety of organic nitrogen compounds. A sample of water, collected in July when the concentration of organic nitrogen compounds in solution is at a maximum, was found by this method of analysis to contain 10 mg. N per m.³, and by bio-assays with two species of phytoplankton to contain 11·5 and 13 mg. N per m.³ available to the plants (*J. Mar. Biol. Ass. U.K.* **36**, 157, 1957).

A method for the determination of nitrate in sea water by reduction to nitrite, by phenyl hydrazine in alkaline solution in the presence of copper, has been described by Mullin and Riley (*Analyt. chim. acta,* **12**, 464, 1955). Under the specified conditions reproducible yields of c. 85% were obtained with a standard deviation of 2%. Nitrite if present in the sample is partially destroyed, but this can be allowed for.

The following books have been recently published:

DIETRICH, G. and KALLE, K. *Allgemeine Meereskunde,* Berlin, 1957.

HARDY, A. C. *The Open Sea: The World of Plankton.* London, 1956.

MARSHALL, S. M. and ORR, A. P. *The Biology of a Marine Copepod.* Edinburgh, 1955.

VINOGRADOV, A. P. *The Elementary Chemical Composition of Marine Organisms.* New Haven, 1953.

CONVERSION FACTORS

1 knot $= 51\cdot479$ cm. per sec.

1 fathom $= 6$ feet $= 1\cdot8288$ metres.

1 nautical mile $= 1853\cdot25$ metres $= 2026\cdot7$ yards $= 1\cdot1516$ statute miles.

1 acre $= 4047$ square metres.

1 cubic metre $= 1000$ litres.

1 foot-candle $= 10\cdot764$ metre-candles or lux.

1 kilolux of sunlight represents a flow of energy in the visible spectrum (380–720 mμ) equal to:

$0\cdot006$ g. cal. per cm.^{2} per minute

10^{-4} g. cal. per cm.^{2} per second

4186 ergs. per cm.^{2} per second

$1\cdot5$ joules per cm.^{2} per hour

1 quantum of light energy $= \dfrac{47\cdot5 \times 10^{-17}}{A^\circ}$ gram-calories, where A° equals the wave-length of the light in Ångstrom units of 10^{-8} cm.

1 milligram (mg.) per cubic-metre (m.^{3}) $= 1$ microgram (μg or γ) per litre.

1 milligram-atom of an element (mg.-at.) $=$ weight of the element in milligrams divided by its atomic weight, and contains 6×10^{20} atoms.

1 milligram-atom per cubic metre $= 6 \times 10^{11}$ atoms per mm.^{3}

The concentration of dissolved phosphate is expressed in different units, in the extensive literature concerning its distribution.

mg. P per m.^{3}	2	4	6	8	10	12	14	16
mg.-atoms P per m.^{3}	0·06	0·13	0·19	0·26	0·32	0·39	0·45	0·52
mg. P$_2$O$_5$ per m.^{3}	5	9	14	18	23	28	32	37
mg. PO$_4$ per m.^{3}	6	12	18	24	31	37	43	49
mg. P per m.^{3}	18	20	22	24	26	28	30	32
mg.-atoms P per m.^{3}	0·58	0·64	0·71	0·77	0·84	0·90	0·97	1·06
mg. P$_2$O$_5$ per m.^{3}	41	46	50	55	60	64	69	73
mg. PO$_4$ per m.^{3}	55	61	67	73	80	86	92	98

Caution. The concentrations, obtained by visual comparison, given in the older literature have in many cases not been corrected for 'salt error', the 'corrected' values are then about $1\cdot3$ times the uncorrected values (Cooper, 1938).

INDEX

Phytoplankton (*cont.*)
  respiration, 87, 89, 95
  buoyancy and sinking, 102
  'physiological state' of, 102, 111
  culture of, 103–13
  succession of species, 103
  and silicate supply, 105, 109
  and B_{12} (cobalamin), 107
  and divalent sulphur compounds,
    107–9
  and bacterial flora, 112
Plant pigments, 90
Porcupine Bank, 100
Potassium, 4, 30, 131, 135
Pressure
  effect on pH, 156
  effect on CO_2 dissociation, 180
Proteins
  in unicellular algae, 28
  in solution, 151
Prymnesium parvum, 112
Pteropod ooze, 29

Radiolarian ooze, 29
Radium, 141
Refractive index of sea waters, 130
River waters, 42, 54
Rubidium, 141
Respiration
  bacterial, 69, 71
  phytoplankton, 87, 89, 95
  animal, 116–20

Salinity, 3, 125
  distribution in Atlantic Ocean, 11,
    15
  relation to density, to 'chlorinity',
    to freezing-point, to osmotic and
    vapour pressure, to viscosity and
    refractive index, 128–30
Salt error
  of indicators, 157
  in visual estimation of phosphate,
    217
Sargasso Sea, zooplankton population,
  121
Scandium, 141
Scattering layer, 122
Scotia Sea, 98
Sea water, artificial, 113, 137
Sea weeds, trace-elements in, 29,
  139
Secchi disk, 30, 86
Selenium, 140, 141
Shell deposits, phosphate in, 42

Silica
  in deposits, 29
  resolution from diatoms, 40
  and diatom growth, 105–9
  in solution, distribution, 147–9
  estimation, 198
Silver, 141
Skeletonema costalum, 104–9
Sodium, 4, 132
Soil extract in culture of phytoplankton, 105
Solubility
  copper, 141
  ferric hydroxide, 142
  carbon dioxide, 168
  oxygen, 184
  nitrogen, 185
  argon, 186
Solubility product
  magnesium carbonate and hy-
    droxide, 132
  calcium carbonate, 133
Specific alkalinity, 161
Specific gravity
  and salinity, 128
  and temperature and pressure,
    130
Standard sea water, 126
Storage of sea water
  bacterial changes, 64
  chemical changes, 69
  for subsequent analysis, 192, 194
Strontium, 4, 135
Submarine ridges and currents, 16
Sulphates, 4, 136
Sulphides, 35, 108
Sulphur compounds and diatom
  growth, 108
Sugar
  isotonic with sea water, 3
  and pumping by oysters, 150
Surface active compounds, 67, 151

Tees Estuary, 21
Temperature
  Atlantic Ocean, 14
  effect on algal respiration, 94; on
    photosynthesis, 80, 94; on meta-
    bolic rate of animals, 119
Thermocline, 16–18, 100
Thiamin
  and algal growth, 107
  in lake water, 151
Thiourea, 108
Thorium, 141, 194